Microgrids and Local Energy Systems

Edited by Nick Jenkins

Published in London, United Kingdom

IntechOpen

Supporting open minds since 2005

Microgrids and Local Energy Systems
http://dx.doi.org/10.5772/intechopen.90649
Edited by Nick Jenkins

Contributors
Ali Al-Wakeel, Meysam Qadrdan, William Seward, Weiqi Hua, Janake Ekanayake, Yue Zhou, Jianzhong Wu, Wenlong Ming, Ryan Merlin Yonk, Corbin Clark, Jessica Rood, Carlos E. Ugalde-Loo, Ivan De la Cruz

First published in London, United Kingdom, 2021 by IntechOpen
IntechOpen is the global imprint of INTECHOPEN LIMITED, registered in England and Wales, registration number: 11086078, 5 Princes Gate Court, London, SW7 2QJ, United Kingdom
Printed in Croatia

British Library Cataloguing-in-Publication Data
A catalogue record for this book is available from the British Library

Additional hard and PDF copies can be obtained from orders@intechopen.com

Microgrids and Local Energy Systems
Edited by Nick Jenkins
p. cm.
Print ISBN 978-1-78984-530-3
Online ISBN 978-1-78984-531-0
eBook (PDF) ISBN 978-1-78984-556-3

An electronic version of this book is freely available, thanks to the support of libraries working with Knowledge Unlatched. KU is a collaborative initiative designed to make high quality books Open Access for the public good. More information about the initiative and links to the Open Access version can be found at www.knowledgeunlatched.org

For EU product safety concerns:
IN TECH d.o.o., Prolaz Marije Krucifikse Kozulić 3, 51000 Rijeka, Croatia, info@intechopen.com or visit our website at intechopen.com.

Meet the editor

Nick Jenkins leads the Centre for Integrated Renewable Electricity Generation and Supply (CIREGS) at Cardiff University, Wales. Before moving to academia, his career included fourteen years of industrial experience, five of which were in developing countries. When in industry, Mr. Jenkins worked on electric power systems as well as for manufacturers of wind turbines and photovoltaic equipment. While at the university, he has developed teaching and research activities in electrical power engineering and renewable energy. He is a fellow of the Royal Academy of Engineering, Institution of Engineering and Technology (IET), Institute of Electrical and Electronics Engineers (IEEE), and Learning Society of Wales.

Contents

Preface **XI**

Section 1
Microgrids **1**

Chapter 1 **3**
Protection of Microgrids
by Janaka Ekanayake

Chapter 2 **37**
Power Electronic Converters for Microgrids
by Wenlong Ming

Chapter 3 **65**
Peer-to-Peer Energy Trading in Microgrids and Local Energy Systems
by Yue Zhou and Jianzhong Wu

Section 2
Local Energy Systems **89**

Chapter 4 **91**
District Heating and Cooling Systems
by Iván De la Cruz and Carlos E. Ugalde-Loo

Chapter 5 **127**
Electricity Storage in Local Energy Systems
by William Seward, Weiqi Hua and Meysam Qadrdan

Chapter 6 **153**
Local Energy Systems in Iraq: Neighbourhood Diesel Generators and Solar Photovoltaic Generation
by Ali Al-Wakeel

Chapter 7 **179**
Regulatory Impediments to Micro-Wind Generation
by Ryan M. Yonk, Corbin Clark and Jessica Rood

Preface

Energy systems throughout the world are in a period of rapid change. Unfortunately, energy systems are a major contributor to gaseous emissions through the combustion of fossil fuels. Levels of greenhouse gases in the atmosphere must be controlled if climate change is to be mitigated. Almost one billion people are without any access to electricity and many more have a power supply that is frequently interrupted, often for hours at a time. Any discussion of energy systems must address more than electricity with the energy required for heating and transport exceeding that used for generating electricity in many countries. It is estimated that around three billion people do not have access to clean energy for cooking and rely on open fires and traditional biomass, with damaging consequences for their health and the local ecosystems.

When public supplies of gas and electricity gas were first introduced more than 100 years ago, town gas was made from coal in local gasworks and electricity was generated in small power stations that served only a few square kilometres. The operation and management of these historic local energy systems was the responsibility of the local town or city council. Over time and as the technology improved, larger and more powerful electricity-generating stations and sources of gas were needed, and interconnected gas and electricity pipe and cable networks now cover large areas of many countries. The local gasworks were shut down and replaced with supplies of natural gas piped from remote gas wells and electricity generating stations were moved out of the cities to allow larger generators to be used and to improve urban air quality. The concept of having large central power stations located far away from centres of population and a strong transmission grid together with extensive distribution networks has dominated electricity supply since 1950. A similar approach has been adopted for the supply of gas. Bulk supplies of natural gas are transported by pipeline or in ships, and high-pressure transmission pipes feed lower-pressure distribution networks. Large organisations are needed to manage and finance these large complex systems and thus strong gas and electricity utilities, often publicly owned, have emerged.

Over the last 10–15 years, the possibility of reverting, at least in part, to smaller, local energy systems has become an important topic of research in many countries. This change of emphasis has been driven by multiple factors, including technical, financial, and societal pressures. These pressures vary country by country but are already changing the ways that energy consumers behave and how energy grids are developed and managed.

In a low-carbon energy system, electricity generation relies increasingly on renewable energy sources, particularly the sun and wind. These energy sources are diffuse and dispersed over wide areas, and the wind turbines or solar PV systems are connected to the local distribution networks to collect the power. Renewable sources of energy are only generated when the sun or the wind are suitable and thus do not have the important attribute of fossil fuels of naturally storing large quantities of energy in chemical form. The demand and supply of an electricity system must always balance, and for an electricity system powered by renewables, this is a major challenge. The cost of battery energy storage remains and so it is hard to see how costs of the transition to a low-carbon electricity system can be controlled unless customer load is closely integrated into its operation. The engagement of customers with their energy supply then becomes very important and helps explain the current interest in microgrids and local energy systems.

It is cheaper to store energy as heat or in compressed gas than as electricity, and the integration of heat and gas systems with electricity networks is a promising avenue of research. This is often described as providing flexibility to the electricity system and can be arranged conveniently at a local level. It is a key attribute of local energy systems.

The book describes recent research undertaken, mainly at Cardiff University, to address important topical questions of microgrids and local energy systems. It provides up-to-date, new information in an accessible form as well as comprehensive references.

The book is divided into two sections. The first section "Microgrids" begins with a chapter reviewing the electrical protection of microgrids using several case studies. This chapter is followed by a study of the power converters used and the use of multi-objective optimization for the selection of component ratings. This section of the book concludes with a chapter on the use of peer-to-peer energy trading in microgrids and again uses case studies to demonstrate the techniques.

The second section of the book addresses local energy systems, including heat and gas networks as well as electrical systems. The first chapter in this section discusses district heating and cooling systems and their control. The second chapter in this section addresses the important question of energy storage. This is followed by a description of the local neighbourhood generators used to supplement the utility electricity supply in Iraq and some adjacent countries. A final chapter discusses regulatory impediments to micro-wind generation in the United States.

I express my sincere thanks to all the chapter authors, particularly the early career researchers who played such an important role in writing some chapters. I am also grateful to the staff at IntechOpen.

Nick Jenkins
Department of Engineering,
Cardiff University,
Cardiff, UK

Section 1

Microgrids

Chapter 1

Protection of Microgrids

Janaka Ekanayake

Abstract

The concept of microgrids goes back to the early years of the electricity industry although the systems then were not formally called microgrids. Today, two types of microgrids can be seen: independent and grid connected. The protection requirement of these two types differs as the protection needs of an independent microgrid are intended for protecting components and systems within the microgrid, whereas a grid connected microgrid demands both internal and external protection. The first part of this chapter is dedicated to independent microgrids. How protection devices such as residual current circuit breakers, miniature and moulded case circuit breakers, and surge protective devices should be selected for an example microgrid is discussed while referring to the relevant standards. In the next section, the protection of a grid connected microgrid is discussed. Particularly, micro-source protection, microgrid protection, loss of mains protection and fault ride-through requirements are discussed while referring to two commonly used distributed generator connection codes. An example with simulations carried out in the IPSA simulation platform was used to explain different protection requirements and calculation procedures. Finally, grounding requirements are discussed while referring to different interfacing transformer connections and voltage source inverter connections.

Keywords: microgrid, micro-sources, protection, grid connected, independent, islanding

1. Introduction

Remote locations which are far from grid electricity and islands have developed independent grids to supply power to the local populations. Even though they were not formally called microgrids, they met most of the criteria for today's microgrid systems. With the advancement of renewable energy technologies and energy storage, the concept of microgrid emerged mainly for grid connected systems that are controlled and operated as a smaller grid.

WG6.22 of CIGRE [1] defines microgrids as "electricity distribution systems containing loads and distributed energy resources (such as distributed generators, storage devices, or controllable loads), that can be operated in a controlled, coordinated way either while connected to the main power network or while isolated." A microgrid can offer a number of benefits to those connected to the microgrid as well as to the local utility. As micro-sources are connected closer to loads, the distribution network losses are reduced considerably. The optimum use of Distributed Energy Resources (DERs) within a microgrid further increases the efficiency of their operation. Further, microgrid connected facilities can continue in operation during a grid outage thus increasing the reliability and quality of supply. They can relieve grid congestion and improve operation of the utility grid.

Based on the purpose and regulatory regime in which they operate, microgrids can be categorise as:

a. "Off-grid" or "independent" – a microgrid that is not connected to the utility grid and serves a remote location or an island;

b. "Campus" or "Customer" – a microgrid connected to the local grid, supply power to one or more premises, and maintain some level of service in isolation from the grid; and

c. "Community" or "Utility" – a microgrid integrated into the utility network serving multiple customers within a community.

In this chapter, for the purpose of protection, "independent" (category a) and "grid-connected" (category b or c) will be used. When the microgrid is in "independent" mode, its protection should disconnect the faulty portion of the microgrid with minimum disruption to the loads connected to the microgrid. When a microgrid is in the "grid connected" mode, it should protect microgrid components when a fault is within the microgrid and isolate or provide fault ride through when a fault is in the utility network to which it is connected. Further, the microgrid protection should be coordinated with the utility network protection while having minimum impact on network protection.

2. Fundamental requirements of protection of a microgrid

Protection is installed to detect fault occurrence and isolate the faulted equipment. This is achieved by a fuse or a circuit breaker (CB). When using a fuse and/or CB for protecting a circuit, the following should be considered:

- Selectivity or discrimination: This is the ability of the protection system to disconnect only the faulted section of a plant and to leave the rest of the power system operating.

- Stability: This refers to the requirement of a protection scheme not to operate for remote or "out-of-zone" faults.

- Speed of operation: Fast protection reduces the risk of damage to plant and personnel, but it is more difficult to make fast-acting protection stable for out-of-zone faults and to provide correct discrimination.

- Sensitivity: This is the level of over-current at which the CB will operate.

- Main protection: This is the primary protection system on any circuit or item of the plant.

- Backup protection: This protection is design to operate in the event of failure of the main protection and to cover certain items of plant, which have a low probability of failing. Backup protection is generally slower than main protection and may isolate more than one circuit.

The example circuit shown in **Figure 1** is used to describe the above in more detail. For a fault at the location shown, it should be cleared by fuse F1 not by CB3. This is called ***selectivity***. F1 and CB3 should be able to discriminate whether the fault

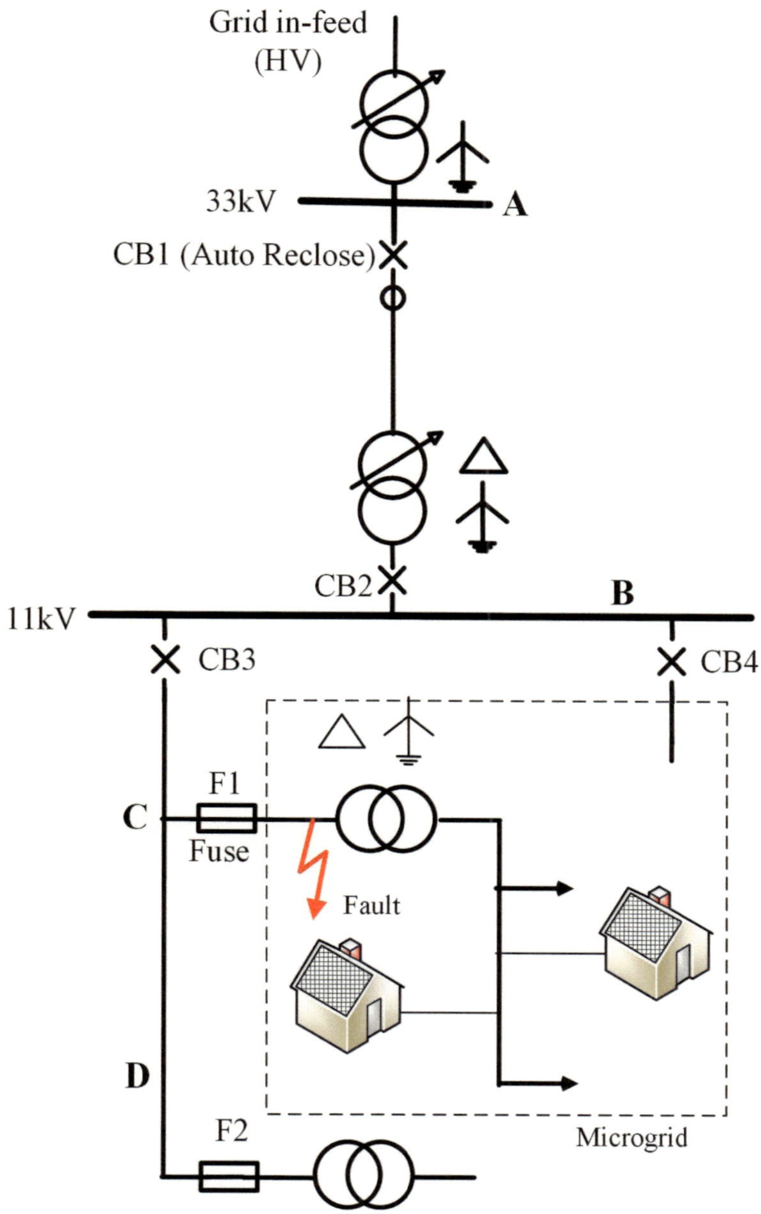

Figure 1.
Grid connected microgrid.

is in its jurisdiction or not. This jurisdiction (or region of operation) of a fuse or a CB is called the ***zone of protection***. CB3 should be able to detect that the fault shown in the figure is ***out-of-zone***. Further, CB3 should provide ***backup protection***, in case F1 is unable to clear the fault.

3. Fault current contribution of different micro-sources and implications for protection

Microgrids utilise hybrid energy sources consisting of renewable energy sources and conventional power plants such as combined heat and power and diesel gensets. In contrast, conventional power plants usually employ synchronous generators.

During a fault, as shown in **Figure 2**, if the microgrid is connected to the utility grid, then there will be fault currents from the utility grid and also from the micro-sources (marked in red lines with arrows). In an independent microgrid that is not connected to the utility grid then the only sources of fault current will be the micro-sources.

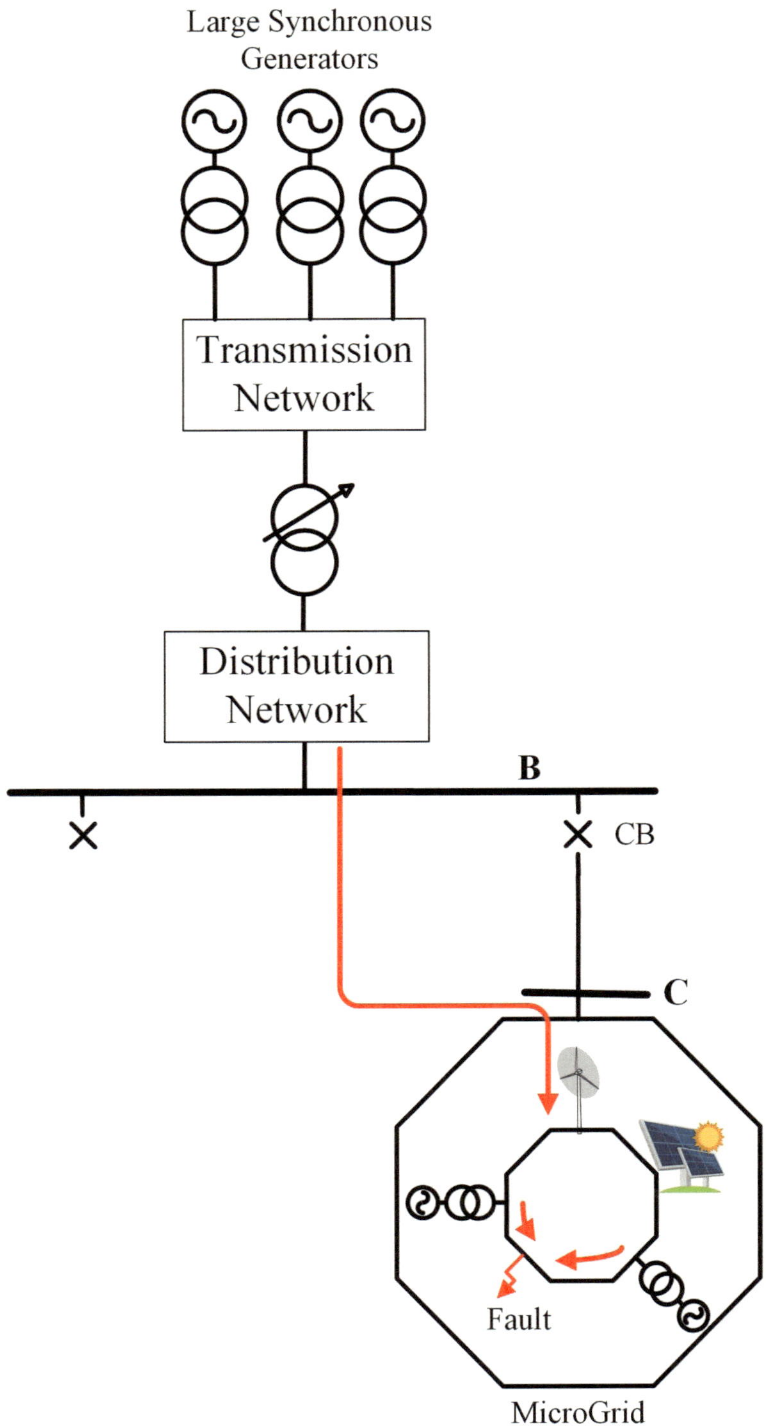

Figure 2.
Fault current within a grid-connected microgrid.

3.1 Fault current contribution from synchronous generators

The fault current contribution from a synchronous generator decreases exponentially after the fault occurs and settles down to a steady-state level. The fault current of a synchronous generator into a three-phase fault depends on the rotor construction. For a cylindrical pole machine, the direct axis has the same value of reactances as the quadrature axis and the fault current contribution to a three-phase fault is usually described by an expression of the form [2–4]:

$$I(t) = \mathrm{E_F}\left[\frac{1}{\mathrm{X}} + \left(\frac{1}{\mathrm{X}'} - \frac{1}{\mathrm{X}}\right)e^{-t/\mathrm{T}'} + \left(\frac{1}{\mathrm{X}''} - \frac{1}{\mathrm{X}'}\right)e^{-t/\mathrm{T}''}\right]\cos(\omega t + \lambda) - \frac{\mathrm{E_F}}{\mathrm{X}''}e^{-t/\mathrm{T_a}}\cos(\lambda) \quad (1)$$

where.

X	synchronous reactance
X'	transient reactance
X''	sub-transient reactance
E_F	pre-fault internal voltage
T'	transient short circuit time constant
T''	sub-transient short circuit time constant
T_a	armature (dc) time constant
λ	angle of the phase at time zero
ω	system angular velocity

The armature time constant (T_a) is not a fixed value but depends on the location of the fault. It is given by

$$\mathrm{T_a} = \frac{(\mathrm{X}'' + \mathrm{X_e})}{\omega(\mathrm{R_a} + \mathrm{R_e})} \quad (2)$$

where.

X_e	external reactance (to the fault)
R_e	external resistance (to the fault)
R_a	armature resistance

The first three terms of Eq. (1) represent a symmetrical decaying ac fault current and the fourth term a dc offset.

Table 1 shows the parameters of a large (G_L) and a small (G_S) synchronous generator. The main difference is the resistance of G_S is much higher than that of G_L and the time constants of G_S is much lower than G_L. For a three-phase fault at the terminal of the machine, the armature time constant of G_L is 163 ms and that of G_S is 5.9 ms. **Figure 3** shows the envelope of the current immediately after the terminal fault for both machines. As can be seen G_S has a shorter transient compared to G_L. However, as G_L is connected far away from the microgrid, its contribution for a three-phase fault within the microgrid is much smaller and the presence of R_e and X_e minimises the armature time constant and the period of transient.

Parameter	Larger generator (G_L)	Smaller generator (G_S) [5]
Rated voltage	15 kV	400 V
Rated Apparent Power	21.4 MVA	16 kVA
Base impedance (Ω)	10.5	10
X (pu)	2.04	3.82
X′(pu)	0.36	0.26
X″(pu)	0.22	0.1
R_a (Ω)	0.045	0.54
Field Resistance (Ω)	0.18	9.79
T′	1.4 s	41 ms
T″	18 ms	6 ms

Table 1.
Comparison of parameters of a large and small synchronous generator.

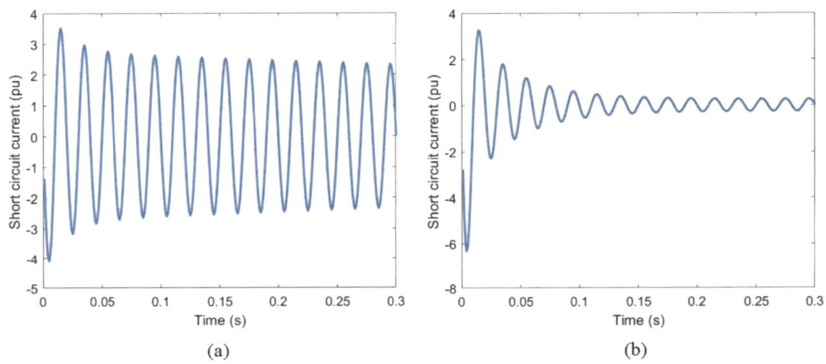

Figure 3.
Fault current at the terminal of the synchronous generator. (a) for G_L and (b) for G_S.

During the fault, the generator terminal voltage decreases rapidly, and as the field current is usually derived from the generator terminal, the field current also decreases. This limited field current capability reduces the airgap flux thus reducing the fault current contribution of the synchronous generator. The fault current will fall to zero within a fraction of a second or within a few seconds at the most, thus generator protection device such as a moulded case circuit breaker (MCCB) may not detect the fault. This fault current without field forcing is shown by the red line in **Figure 4**. For the correct operation of the protective devices, field forcing, which is maintaining the field current throughout a fault condition, is often required. Fault current with field forcing is shown by the blue line. Reference [6] highlights a number of field forcing methods such as shaft mounted permanent magnet generator as a pilot exciter and auxiliary winding design that enables to obtain the field current from both current and voltage of the generator.

3.2 Fault current contribution from asynchronous generators

Fixed speed wind farms and some small hydro plants employ asynchronous (induction) generators as their power source. In these generators, the airgap flux is formed by the induction effect, and when the stator supply is lost, the airgap flux also diminishes. Therefore, as **Figure 5** shows typical fault current shows a high initial current and a rapid decay to zero. Eq. (3) is used to describe the fault current contribution of an asynchronous generator to a three-phase fault at its terminals.

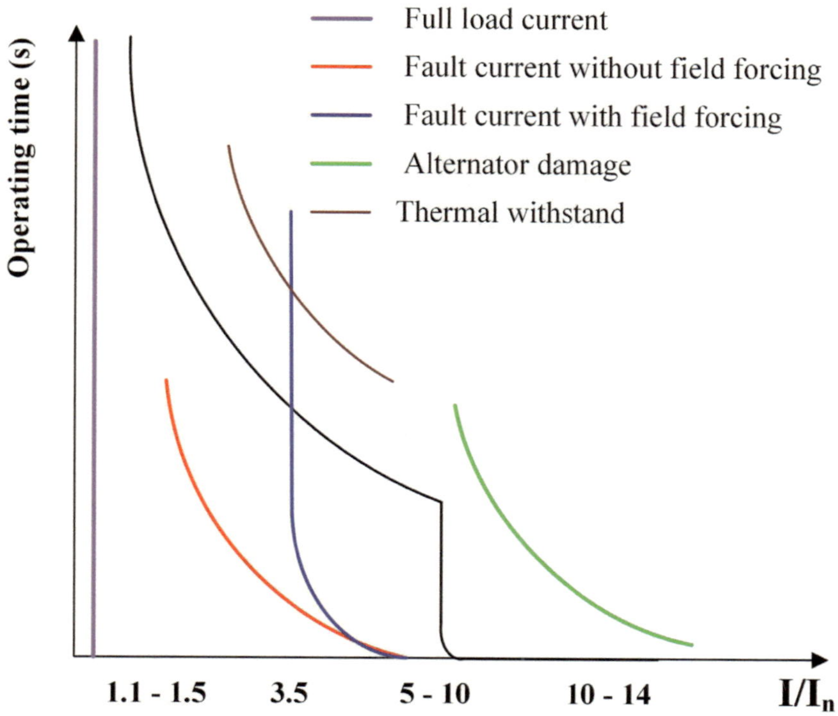

Figure 4.
Characteristic of the generator protection for terminal faults [6].

$$I(t) = \frac{V_1}{X''}\left[\cos(\omega t + \lambda)e^{-t/T''} + \cos(\lambda)e^{-t/T_a}\right] \quad (3)$$

where.

$X'' = X_1 + \frac{X_2' X_m}{X_2' + X_m}$; $T'' = \frac{X''}{\omega R_2'}$; $T_a = \frac{X''}{\omega R_1}$.	
R_1	is the stator resistance.
X_2' and R_2'	are the stator referred rotor reactance and resistance.
X_m	is the magnetising reactance.
V_1	is the magnitude of the network voltage.
$\omega = (1 - s)\omega_s \approx \omega_s$	(as s is very small).

Any external impedance involved must be added to the stator impedance.

3.3 Fault current contribution of converter connected sources

Solar plants and variable speed fully rated wind farms are interfaced to the microgrid through a voltage source converter (VSC). The VSC effectively creates a synthesised ac voltage behind an inductor. The VSC employs an inner control loop which is usually a current controlled regulator. During a network fault, in between the time of disconnecting the upstream protection and islanded detection of VSC, the inner control loop of the VSC regulates the positive-sequence component of its output current to a constant value. Therefore, most short circuit analysis platforms such as ASPEN. CAPE, and ETAP model VSCs as a voltage controlled current source [7].

VSC usually employs Insulated-gate bipolar transistors (IGBTs) and these will be damaged even when short duration high currents flow through them. When a

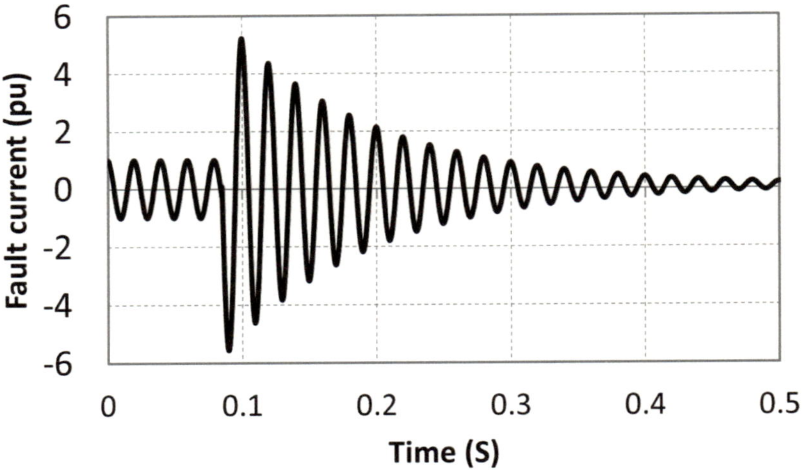

Figure 5.
Fault current of an induction generator with a three-phase fault at its terminals (phase shown with minimum DC offset).

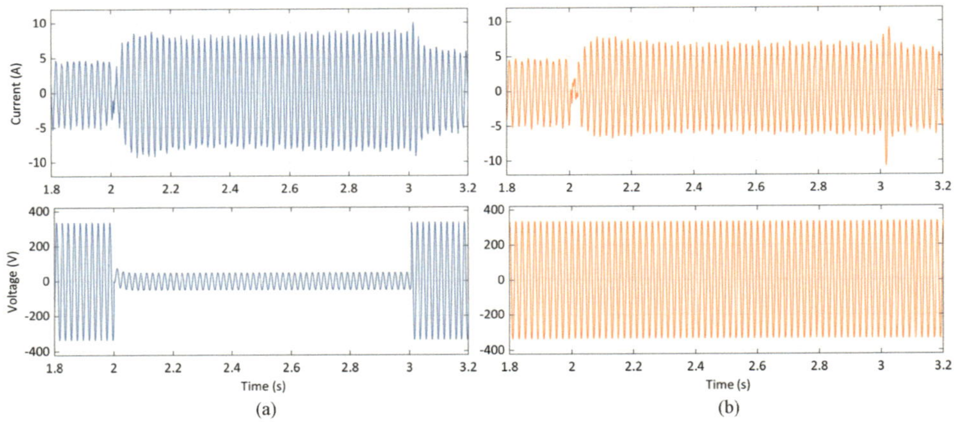

Figure 6.
The solar inverter fault current for a line-to-ground fault (fault is from 2 to 3 sec). (a) Inverter current (A) – Top and terminal voltage (V) – Bottom of the faulty phase. (b) Inverter current (A) – Top and terminal voltage (V) – Bottom of the healthy phase.

transient current approaches a damaging level, the IGBTs are protected by controllers that turn off the IGBTs. Therefore, the fault current contribution from a converter connected distributed generator is very low.

The VSC's response to negative-sequence imbalance, such as a line-to-line or line-to-ground fault, depends on the inverter's control algorithms. Often the VSC provides a constant positive sequence current with very little negative and zero sequence current. This is shown in **Figure 6**, obtained from a line-to-ground fault near a 2 kW solar PV plant (simulated using a MATLAB/SIMULINK model).

4. Protection of independent microgrids

4.1 Protection for safety

When designing and implementing an electrical installation of a microgrid, the safety of the people or livestock occupying that buildings or houses connected to the microgrid, the safety of equipment connected to the microgrid, and the safety of the electrical installation of the microgrid should be considered.

In order to provide protection against electric shock resulted from contact with a conductor which forms part of a circuit and would be expected to be live, basic protection such as the insulation of live parts, the provision of barriers, obstacles or enclosures to prevent touching, and placing out of reach are commonly employed.

4.2 Over-current and short circuit protection

A protective device should automatically interrupt the supply to the line conductor or equipment in the event of a fault. For automatic disconnection, a Residual Current Device (RCD) such as a Residual Current operated Circuit Breaker (RCCB) or an over-current protective device such as fuses, Miniature Circuit Breakers (MCB), Moulded Case Circuit Breakers (MCCB) or Residual Current operated Circuit Breaker with integral Over-current protection (RCBO) could be used.

4.2.1 Commonly used protective devices

4.2.1.1 Fuses

A High Breaking Capacity (HBC) or High Rupturing Capacity (HRC) fuse is used to protect various equipment in a microgrid. When the fault current flows through the fuse, it's fuse wire heats up and melts. The time that corresponds to this process is called pre-arcing time. Even if the fuse wire melts, the current continues to flow in the form of an arc. That time is called the arcing time. In an HBC fuse in order to quench the arc, a powered filling is used, as shown in **Figure 7**.

When selecting a fuse for microgrid equipment, two currents should be considered:

- Rated current: The current that the fusing element can take without melting.

- Breaking capacity: The maximum current the fusing element can break without any damage to the fuse.

Figure 8 shows the time-current characteristic of a fuse. As can be seen, the operating time is inversely proportional to current.

4.2.1.2 Miniature circuit breakers (MCB)

An MCB has a magnetic coil and a bimetal strip (**Figure 9**). For a prolonged over-load current, the bimetal strip starts bending and breaks the circuit after some time. A magnetic coil is an instantaneous element that breaks the circuit immediately after a high current is detected. This acts on a short-circuit current.

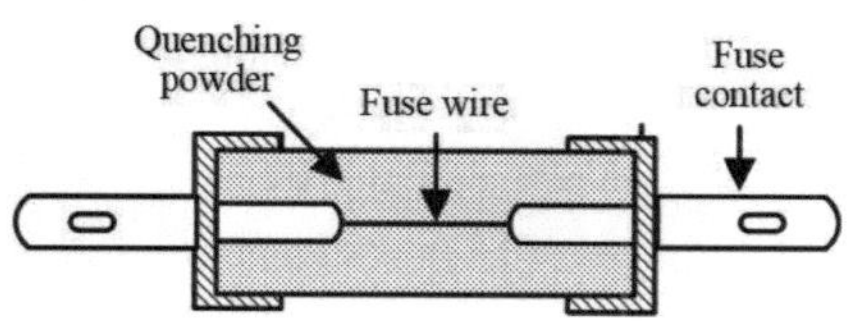

Figure 7.
HBC fuse.

Due to the presence of bi-metal strip and magnetic coil, an MCB shows two different characteristics for over-load and short-circuit currents, as shown in **Figure 10**. In that figure, the PQ characteristics are due to the bi-metal strip and the QR characteristics are due to the magnetic coil.

4.2.1.3 Residual current devices

Residual current devices are used for rapid disconnection of the source of electricity. Their sensitivity can be 10, 30, 100 or 300 mA. **Figure 11** shows part of the operating mechanism in an RCCB. The current in the live wire passes through coil A and that of neutral wire passes through coil B. If phase current and neutral currents are equal, then the flux produced by coils A and B will cancel out. If an earth fault occurs in part of the installation or a person or animal comes into contact with the live wire, the current in one coil differs from that of the other. Then the flux of the two coils will be out of balance resulting in a net magnetic flux in the toroid. This resultant flux links with the fault detection coil and the emf induced will cause the circuit breaker to break the circuit.

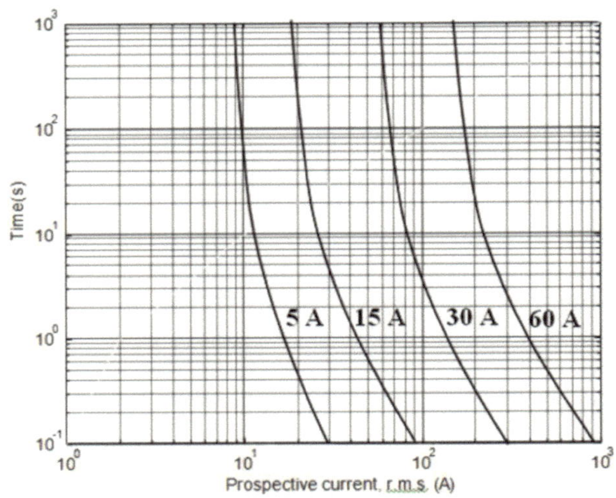

Figure 8.
Operating characteristic of a fuse.

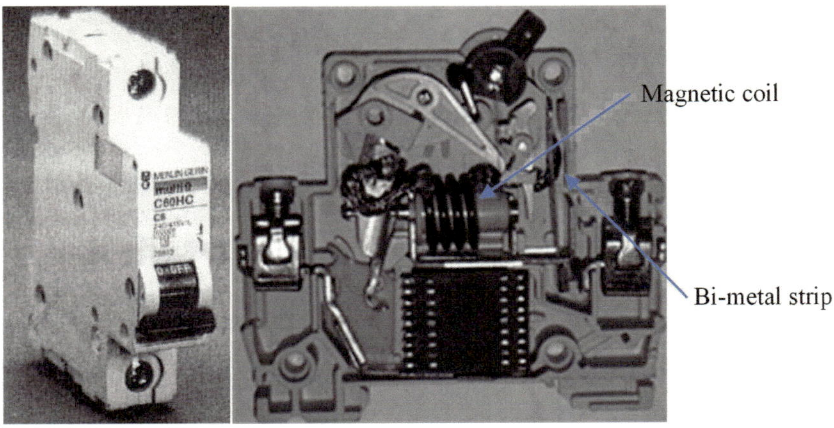

Figure 9.
Open-up section of an MCB.

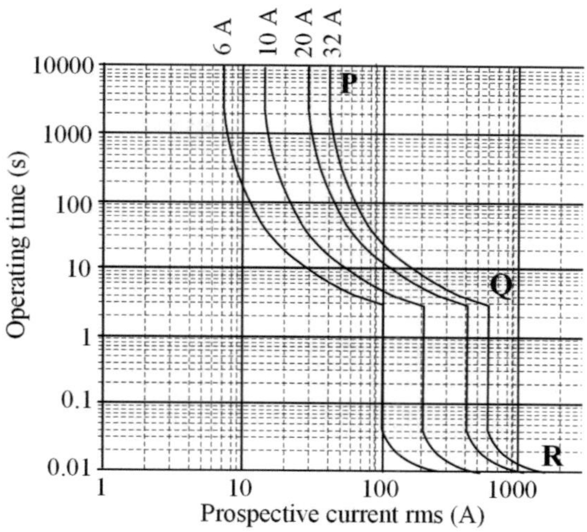

Figure 10.
MCB characteristics.

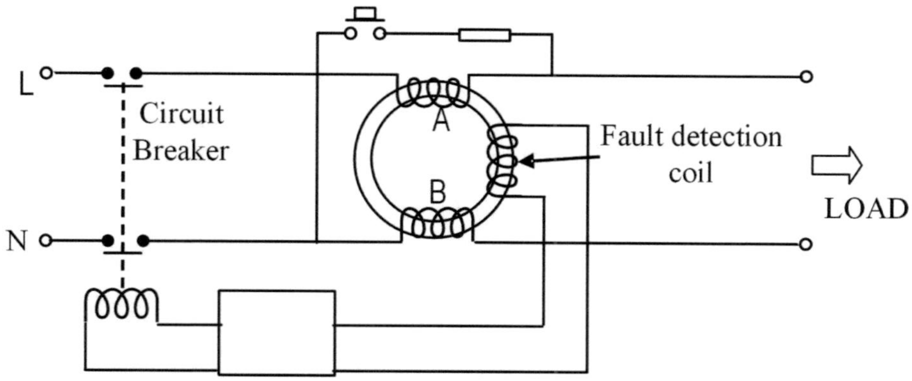

Figure 11.
Simplified diagram of an RCCB.

4.2.1.4 Moulded case circuit breaker (MCCB)

Functionally an MCCB is very similar to an MCB, except the rated current and breaking capacity is much higher. MCCBs are available in three types: Type B, C and D. Type B is used when the power factor of the equipment being protected is closer to 1. Type C is used for inductive loads and Type D is used for motor loads. **Figure 12** shows the symbol of an MCCB and its time-current characteristics. As indicated in the symbol, the MCCB has a circuit breaking element ( ), a thermal element for over-current protection ( ), and a magnetic element for short circuit protection ( ). A modern MCCBs comes in various frame sizes, and the current a frame can normally carry is the rated current of the MCCB.

4.2.2 Protection of solar plants

Figure 13 shows a typical medium-size solar PV plant. The dc circuits are protected by array fuses, and string fuses and disconnectors. A string fuse with a disconnector should be employed for each string. As the fuses used in the strings are subjected to constant sun irradiance and high temperatures, they should be de-rated in many applications by applying the temperature correction factor specified by the

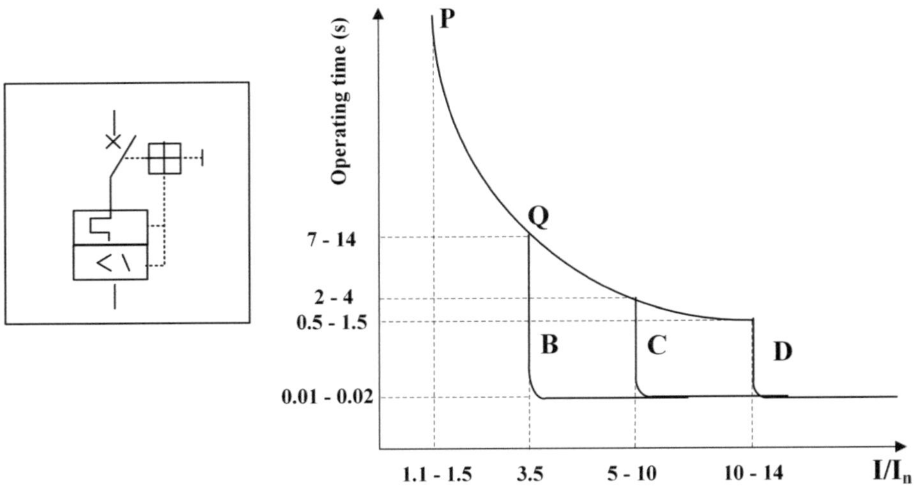

Figure 12.
Symbol of an MCCB and its time-current characteristics.

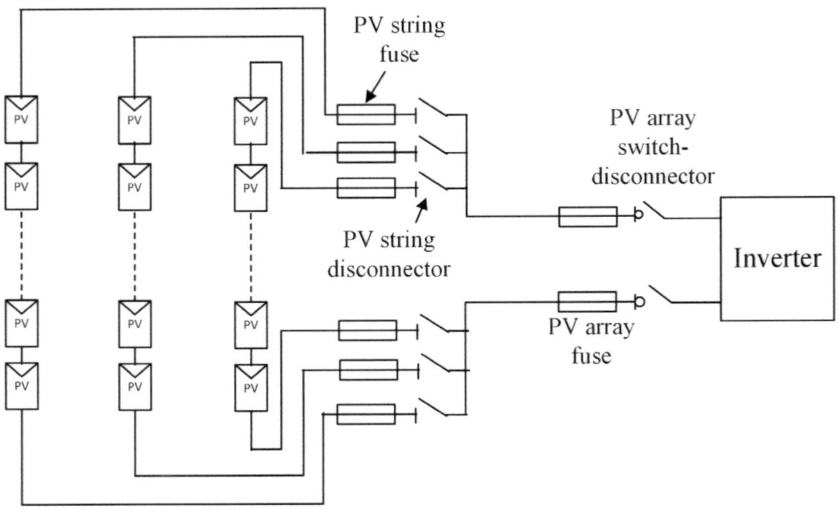

Figure 13.
Typical protection arrangement for a PV plant.

manufacturers. A PV array fuse is only required if there is another source of short circuit current such as a battery system.

IEC 62548 [8] and IEC60364–7-712 [9] provide guidance for protection against over-currents in solar plants. When there are more than two strings in a PV plant, a fuse is required to protect a string from reverse currents due to a short circuit of part of that string. The above IECs state that if $(N_s - 1) \times I_{SC}$ (where N_s is the number of parallel strings and I_{SC} is the short circuit current of a PV sub-array at standard test conditions) is greater than the series fuse rating specified in the manufacturer's data sheet (as per IEC 61730–2 [10]), then string fuses should be employed. String fuses should be selected such that $1.5 \times I_{SC} < I_n < 2.4 \times I_{SC}$ and $I_n \leq$ string fuse rating (where I_n is the rating of the fuse).

4.2.3 Protection of wind power plants

The electrical protection of small wind turbine generators is addressed in IEC61400–1 [11]. The wind turbine generator should have a disconnection switch and the rated current of fuses or the setting current of other over-current protective devices such as MCB and MCCB should be selected as low as possible but adequate for starting the wind turbine generator. Further, the rated short-circuit breaking capacity should be at least equal to the prospective fault current at the point of installation. The action of the bi-metallic strip in MCBs and MCCBs should introduce a time delay that affords the generator time to start and settle into the normal running current without unnecessary tripping.

4.3 Surge protection

Lightning or switching surges can enter into different expensive equipment in a microgrid and destroy or cause it to mal-function. To provide protection against surges, surge protection devices (SPD) are employed. An SPD is a metal oxide varistor or an avalanche breakdown diode or a gas discharge tube. An SPD diverts the surge into the ground, thus providing protection to the device to which it is connected.

4.4 Case study

An islanded microgrid installed in the north of Sri Lanka that feeds power to a remote island that has 120 families is considered. The microgrid has a wind farm, ground-mounted solar plant, battery bank with a grid forming inverter and a backup diesel generator. Some photos from the microgrid are given in **Figure 14** and the rating of these components is given in **Table 2**.

A simplified diagram was created from the single line diagram of Eluvithive microgrid and **Figure 15** shows this simplified network and associated protective devices.

In the event of a network fault, the dc link voltage of the wind turbine will rise rapidly because the inverter is prevented from transforming all the active power coming from the wind generator. Therefore, in order to maintain the dc link voltage below its upper limit, the excess power is dissipated in a chopper circuit shown. Here the excess energy is dissipated into a resistor connected to the dc link as heat. When the dc-link voltage reaches a critical value, the dc-link is short circuited for a short period of time through the chopper. During the activation of the chopper, the dc-link discharges keeping the dc-link voltage below a critical value.

As can be seen from the figure, the wind turbine is protected by a 3 pole over-current protective device (MCCB 1). Each wind turbine is rated at 220 V, 50 Hz but operating at a variable voltage and variable frequency mode. According to the wind turbine specification (WINDSPOT 3.5 kW [12]), the power generation at the cutout wind speed (25 m/s) is 4214 W. The corresponding current is $4214/\left(\sqrt{3} \times 220\right) =$ 11.06 A. The current rating of MCCB 1 was selected as 125% of generator rated current, i.e. 13.8A. Even though a 16 A MCB is adequate when considering the load current rating, the wind turbine manufacturer uses a 25 A MCCB. The selection was based on the short circuit capability of MCBs. They have a breaking capacity less than 5 kA and at the terminal of the generator, the short circuit current may be higher than that. Therefore, the smallest MCCB, having a rated current of 25 A and a braking capacity of 10 kA, is employed. Also, note that as the current from the generator has a variable frequency, an MCCB suitable for operating at variable frequency was selected.

Since the grid side of the wind turbine is a single phase, the maximum current on that side is 19.15 A, thus requiring an overcurrent protective device of 23.9 A (1.25 x 19.15). Therefore, an MCCB having a rated current of 25 A and a braking capacity of 10 kA was used for MCCB 2.

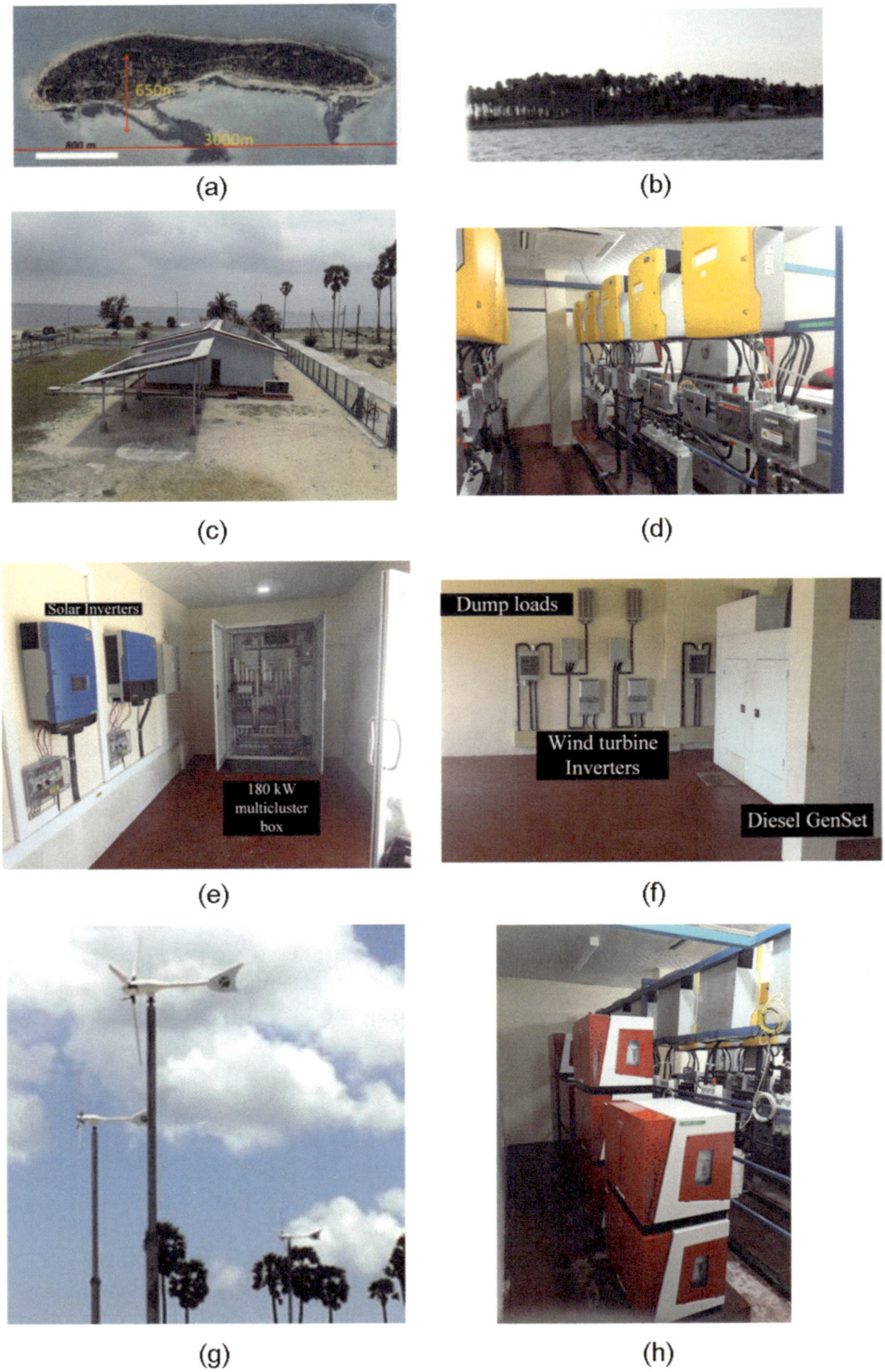

Figure 14.
Some photos from Eluvithive microgrid. (a) Map of the island, (b) distance view of the island, (c) solar panel and microgrid assembly, (d) solar inverters, (e) solar inverters and multi-cluster panel, (f) diesel generator, wind turbine inverters and dump loads, (g) wind turbines, and (h) Li-ion batteries.

Table 3 provides the specifications of the solar panels that are essential for determining the protective devices. In that table STC stands for standard test condition, i.e. irradiance of 1000 W/m^2 and ambient temperature of 25°C.

As there are three parallel-strings, as described in Section 3.2.2, $(N_s - 1) \times I_{SC}$ = 2 x 8.81 = 17.62 A. As this is greater than PV module's series fuse rating, 15 A, string fuses are required. The current rating of the fuse should be such that $13.2\,(1.5 \times 8.81) < I_n < 21.1\,(2.4 \times 8.81)$. Therefore 15 A fuses are employed.

The output current of the inverter is $15 \times 10^3/(\sqrt{3} \times 400) = 21.6$ A, and the MCCB rating was selected as greater than 125% of the output current, i.e. 27 A. Therefore, for MCCB 3, 32 A, 4 pole, 10 kA, MCCB was used. A similar procedure was used to select MCCBs for other places.

Source	Unit capacity	Number of units	Total capacity
Diesel Generator	30 kW	1	30 kW
Wind Plant	3.5 kW	6	21 kW
PV plant	250 Wp	180	45 kWp
Li-ion batteries	13 kW/27.5 kWh	4	52 kW/110 kWh

Table 2.
Different sources and capacity of microgrid components.

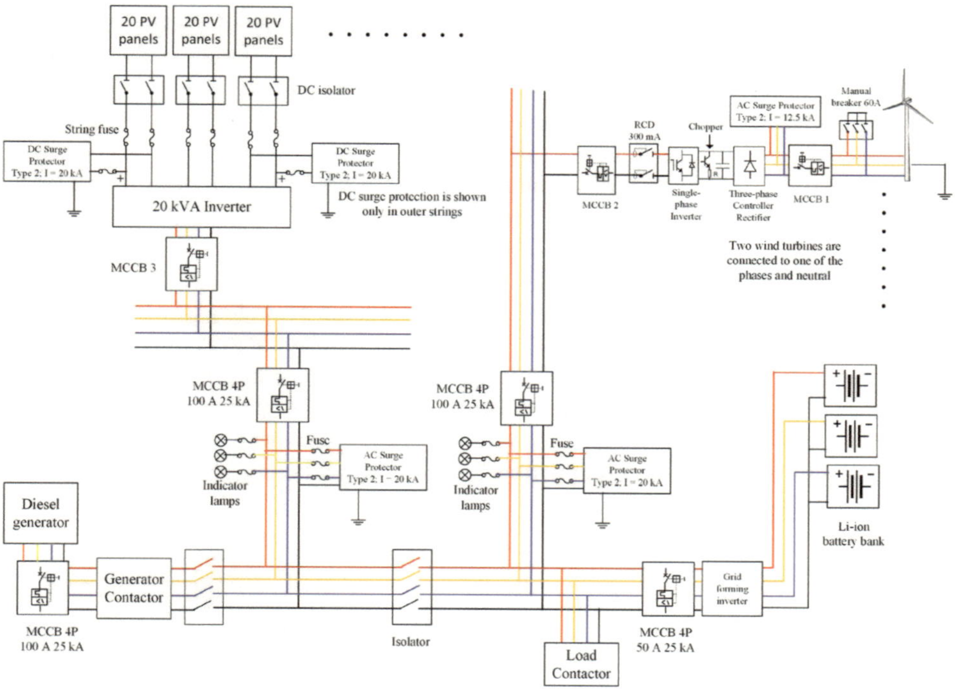

Figure 15.
Example isolated microgrid.

STC Power (Wp)	250
STC Open circuit voltage Voc (V)	37.6
STC short circuit current Isc (A)	8.81
Series string fuse rating (A)	15

Table 3.
Details of each of the solar panels.

	At junction box		Inverter dc side	Inverter ac side	
Criteria	Ldc < 10 m	Ldc > 10 m		Lac < 10 m	Lac > 10 m
Type of SPD	No need	SPD1 Type 1	SPD2 Type 2	No need	SPD3 Type 2

Table 4.
SPD requirement [7].

As shown in the figure, Surge Protection Devices (SPD) are employed to protect sensitive electrical equipment like the inverter, monitoring devices and PV modules. Their requirement depends on the length of the dc cable between the array and inverter (L_{dc}) and the length of the ac cable between the inverter and main distribution board (L_{ac}). **Table 4** specifies these requirements.

5. Protection of microgrids operating in parallel with the grid

There should be adequate protection to ensure the safe operation of the components within a microgrid and external circuit to which the microgrid is connected. As discussed in Section 3, fuses, MCBs, MCCBs, and RCCBs are used for small microgrids. For large microgrids protection relays and associated circuit breakers are used. A typical microgrid connection with associated protection is shown in **Figure 16**.

In **Figure 16**, a number of fault locations could be identified:

- Fault F1: This is a fault within the micro-source, and the generator protection should clear the fault.

- Fault F2: For a feeder fault within the microgrid, both micro-sources and the main grid will provide fault currents. Fault current provided by MS2 is cleared by the relays associated with CB8. Fault current from the grid and MS1 is cleared by the relays associated with CB6. As the coordination of relays should maintain for grid connected and islanded modes, careful studies should be carried out to ensure proper coordination. Two alternative settings could be used with modern numerical relays, one for the grid connected mode and the other for the islanded mode.

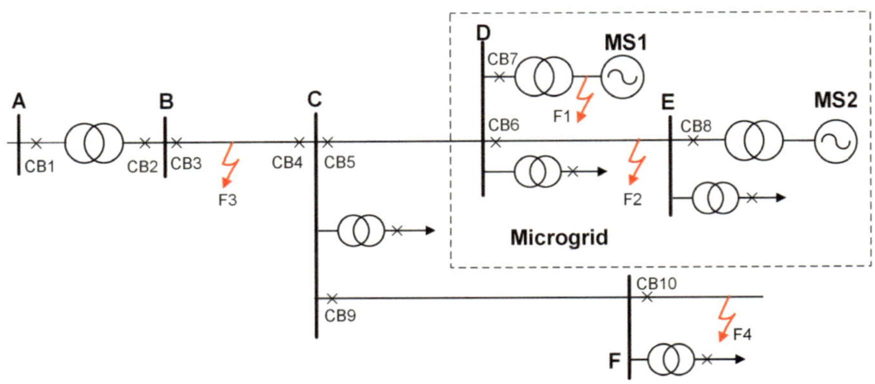

Figure 16.
Protection of grid connected microgrid.

- Fault F3: When a fault occurs in the main grid at the location shown, it is detected by the overcurrent relays associated with CB3 and CB4 and isolates the fault by switching these two circuit breakers. Under such condition, the microgrid experiences a Loss of Mains (LoM) and adequate LoM protection should be employed to isolate the microgrid from the main grid.

- Fault F4: As the fault is at the far end of the radial network, the protection relay CB10 will clear the fault with considerable time delay. During this period the voltages of the busbar C and D may experience voltage dips, and the micro-sources should be able to ride-through the fault and come back to normal operation when CB10 isolates the fault.

The different protection requirements under the above fault conditions are specified in country-specific standards. In the UK, the basic protection requirements for microgrids containing micro-sources of 16 A or less and connected to LV distribution networks are spelt out in the Engineering Recommendation G98 [13]. The protection requirements for a larger microgrid but with less than 50 MW sources and connected to distribution MV networks are given in the Engineering Recommendation G99 [14]. For all distributed energy resources (DER) connected to distribution networks in the USA, the protection requirements are given in IEEE1547: 2018 [15]. G99 specify protection requirements under four generator types classified based on the rated power and the connection point voltage. These categories are given in **Table 5**. IEEE 1547 also categorises generators as category I, II and III based on the abnormal operating performance, mainly voltage and frequency ride-through capabilities. As the recommended protection arrangement varies with the connection arrangement of the transformer and the generator, the readers are directed to the relevant standard for more details.

5.1 Protection of micro-sources

For an internal fault in a large generator, differential protection is usually employed. **Figure 17(a)** shows that if both ends of the stator windings are accessible, differential protection provides phase and earth fault protection. If only one side of the stator is accessible, earth fault protection can be achieved as shown in **Figure 17(b)**. In some cases, differential protection covers both generator and generator transformer. In such a case, a fault F1 shown in **Figure 16** is cleared by the differential protection scheme.

5.2 Protection of the microgrid

For a fault at location F2 of **Figure 16**, micro-sources MS1 and MS2 deliver fault current. As shown in **Figure 18**, an overcurrent relay (marked as 51) associated with the micro-sources can clear the fault. Suppose the micro-source is a small

Type	Output Power	Voltage to which the DER is connected
A	0.8 kW to 1 MW	< 110 kV
B	1 MW to 10 MW	< 110 kV
C	10 MW to 50 MW	< 110 kV
D	> 50 MW	> 110 kW

Table 5.
Different types defined in G99.

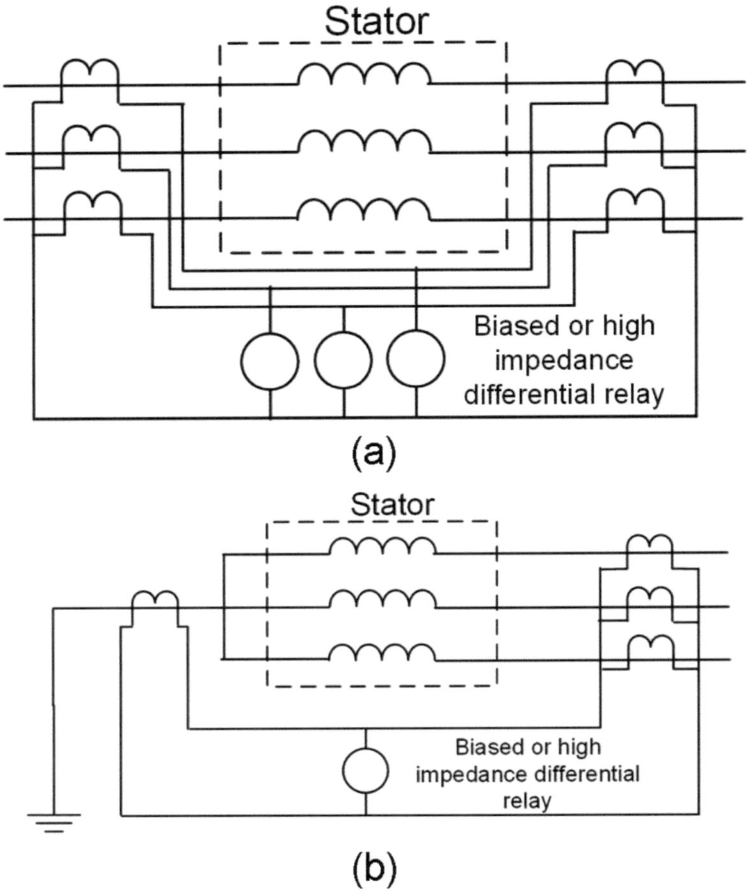

Figure 17.
Differential protection of generator. (a) Earth and phase fault protection and (b) restricted earth fault protection.

synchronous generator. In that case, as discussed in Section 2.1, excitation should be maintained during a fault using a field forcing technique for the over-current protection to work satisfactorily.

A better scheme for providing over-current protection of the network from a small generator is to use a voltage-restrained or voltage-controlled over-current relay (51 V) (also shown in **Figure 18**). This relay requires both current and voltage signals to operate. The voltage restrained approach causes the pick-up current to decrease with decreasing voltage, as shown in **Figure 19(a)**. When the voltage is at its rated value, the pick-up setting of the relay is high. As the voltage decreases (due to a fault), the pick-up value of the relay is decreased proportionally. The voltage-controlled approach has a high pick-up value when the voltage is above a preset voltage, and the pick-up current is reduced to a lower value for voltage below the preset voltage (**Figure 19(b)**).

5.3 Protection for LoM

When the microgrid becomes disconnected from the grid due to an external fault as F3 in **Figure 16**, IEEE 1547 and G99 specify a number of ways to provide LoM protection. Some of the protection functions are associated with generators, whereas some other functions are associated with interface protection associated

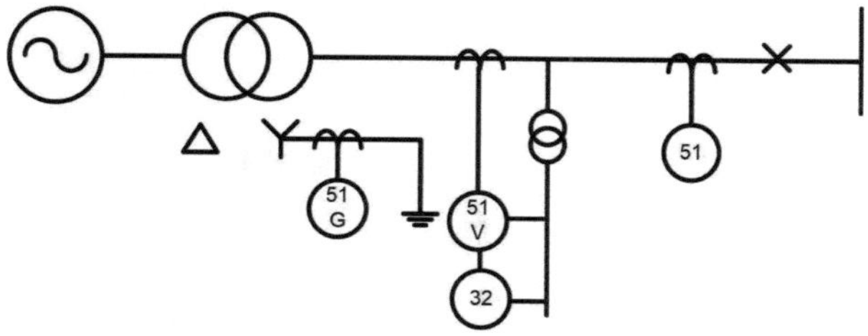

Figure 18.
Overcurrent protection for phase and earth fault protection.

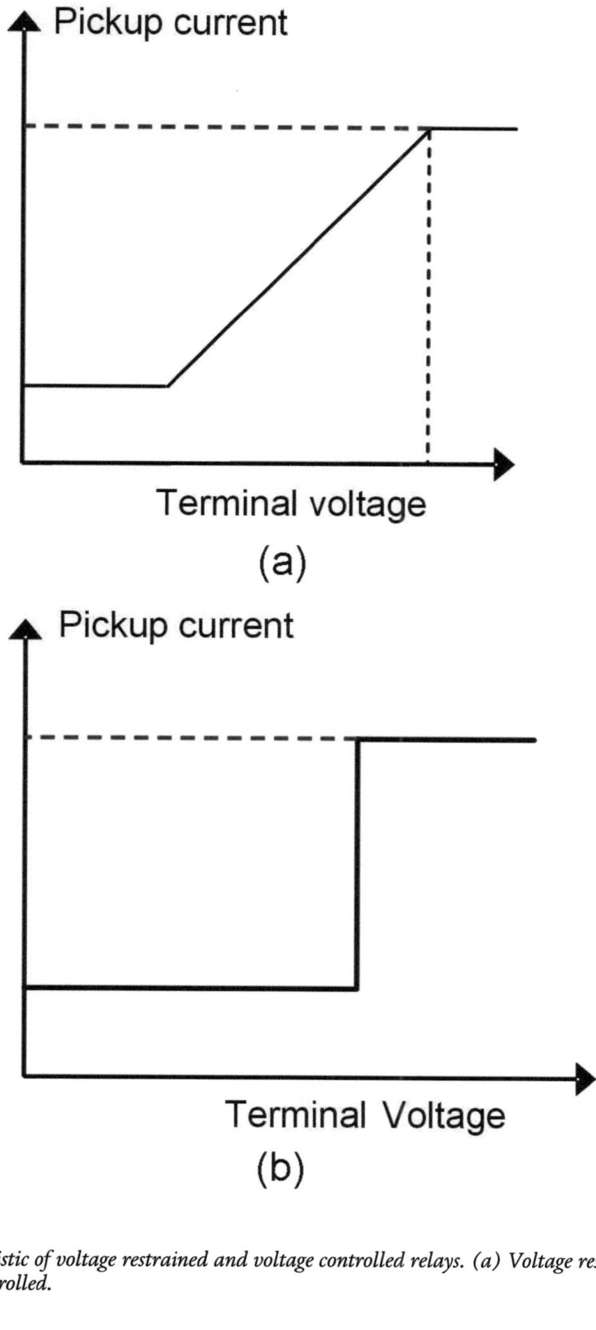

Figure 19.
Operating characteristic of voltage restrained and voltage controlled relays. (a) Voltage restrained and (b) voltage controlled.

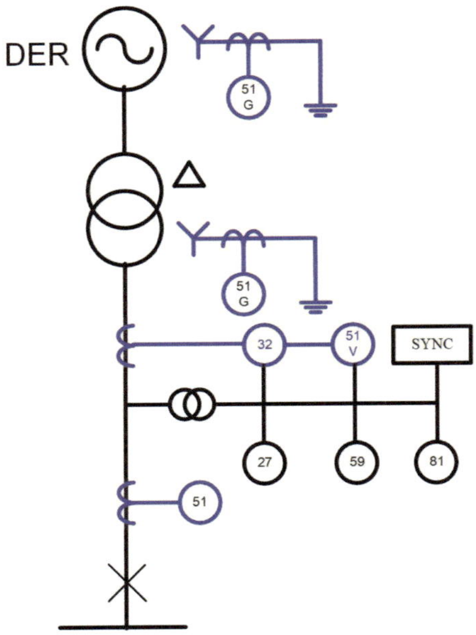

27	Under-voltage
32	Reverse Power
51	Over-current (phase)
51G	Over-current (ground)
51V	Over-current (voltage controlled or voltage restraint)
59	Over-voltage
81	Under and over frequency
SYNC	Synchronising

Figure 20.
Typical protection requirement for DER.

with the utility-tie CB at the point of connection or point of common coupling. **Figure 20** shows the typical LoM protection functions associated with the generator as specified in G99. For completeness, the protection functions described under the previous section are also included in the figure by blue colour lines.

Parameter		IEEE1547:2018			G99	
		Category	Setting (pu & Hz)	Time delay (sec)	Setting (pu & Hz)	Time delay (sec)
Under-voltage (pu)	Stage 1	I	0.70	2.0	0.80	2.5
		II	0.70	10.0		
		III	0.88	21.0		
	Stage 2	I & II	0.45	0.16		
		III	0.50	2.0		
Over-voltage (pu)	Stage 1	I & II	1.10	2.0	LV – 1.14 HV – 1.10	1.0 1.0
		III	1.10	13.0		
	Stage 2	I, II, III	1.20	0.16	LV – 1.19 HV – 1.13	0.5 0.5
Over-frequency (Hz)	Stage 1	I, II, III	62	0.16	52	0.5
	Stage 2	I, II, III	61.2	300		
Under-frequency (Hz)	Stage 1	I, II, III	58.5	300	47.5 Hz	20
	Stage 2	I, II, III	56.5	0.16	47.0	0.5
					LV: 50–1000 V ac LL HV: > 1000 V LL	

Table 6.
Over and under voltage and frequency limits.

Loss of Mains may cause frequency or voltage to be outside the normal operating ranges. When the voltage and frequency are outside the limits defined in **Table 6**, the micro-source is tripped within the respective clearing time. For this, under-voltage (27), over-voltage (59), under-frequency (81/O), and over-frequency (81/U) protection relays are used.

As per the G99, interface protection includes over, and under-frequency relays operating for the 2nd stage settings (**Table 4**), over and under-voltage relays operating for the 2nd stage settings, and a special relay employed for LoM protection.

IEEE 1547: 2018 and G99 state that under and over frequency and under and over voltage relays will not provide adequate protection for LoM, and additional means is required. G99 states that additional LoM protection is required for Type A, B and C power generating modules. Depending on the situations, rate of change of frequency (ROCOF), reverse active power, and reverse reactive power relays may be applied to provide LoM protection. However, if reverse power occurs during normal operation, then the reverse power relay cannot be used for LoM protection. Further, if the microgrid contains micro-sources that can generate reactive power and reverse reactive power flows are experienced during normal operation, then the reverse reactive power relay cannot be used for LoM. Then the ROCOF relay is the only relay that remains for protecting against LoM.

ROCOF relay senses df/dt over a number of cycles and operates when it is greater than a predetermined value. The ROCOF relay should ignore slow changes of df/dt and only should respond to rapid changes. The rate of change of the frequency is determined by the inertia constant[1] of the distributed generator and captive load and governed by the following equations:

$$\frac{d\omega}{dt} = \frac{1}{J}(T_m - T_e) \tag{4}$$

The inertia can be expressed in per unit as an H constant

$$H = \frac{\frac{1}{2}J\omega_S^2}{S_{rated}} \text{ [Ws/VA]} \tag{5}$$

where S_{rated} is the base MVA.
ω_S is the angular velocity (rad/s) at synchronous speed.
Thus

$$\frac{d\omega}{dt} = \frac{\omega^{S}{}_{2}}{2HS_{rated}}(T_m - T_e) = \frac{\omega_s}{2H}(P_m - P_e) \tag{6}$$

where $\frac{\omega_s T}{S_{rated}} = P$ is the per unit (pu) power.
Since $\omega_s = 2\pi f$, from (6) the ROCOF can be obtained as:

$$\frac{df}{dt} = \frac{f}{2H}(P_m - P_e) \tag{7}$$

where P_m and P_e are pu quantities on generator base.

A sudden connection or disconnection of a load and a fault in the network may cause a sudden shift in the terminal voltage vector of a DER with respect to its

---

[1] Inertia constant is the ratio between the kinetic energy and the apparent power of the rotating machine.

normal operating voltage. A relay that measures the voltage phase changes in consecutive cycles and compares the value with a preset value could be used to provide LoM protection. Such relay is called a vector shift relay. Even though the voltage vector shift relay is recognised in G98 as an acceptable method of providing LoM protection, it can be over-sensitive and operate incorrectly. Present revisions of G99 state that the voltage vector shift technique is not an acceptable LoM protection.

5.4 Fault ride-through for remote faults

As discussed before, a remote fault (such as fault F4 in **Figure 16**) may create temporary voltage disturbances. Under such conditions, the micro-sources should ride the fault and resume operate immediately after the remote fault is cleared. As per IEEE 1547, if the temporary voltage is between the lower and upper red lines in **Figure 21**, then the micro-source should ride-through the fault by maintaining synchronism with the grid and without tripping the generator breaker. When the remote fault is cleared, the active current output should be restored to at least 80% of the pre-disturbance active current level within 0.4 s. A similar requirement is specified in G99. The ride-through requirement for G99 Type C & D generators is also shown as the blue line in **Figure 21**.

IEEE 1547 also specifies a rate of change of frequency (ROCOF) ride-through. Micro-sources should ride-through and not trip for frequency excursions having magnitudes of ROCOF less than 0.5 Hz/s for Category I, 2.0 Hz/s for Category II and 3.0 Hz/s for Category III DERs.

5.5 Case study

An example microgrid is shown in **Figure 22**. In that network all the line impedances are given on 200 MVA, 11 kV base. The transformer reactances and generator impedances are given on their own base. This network is used to demonstrate the operation of the microgrid under different faults.

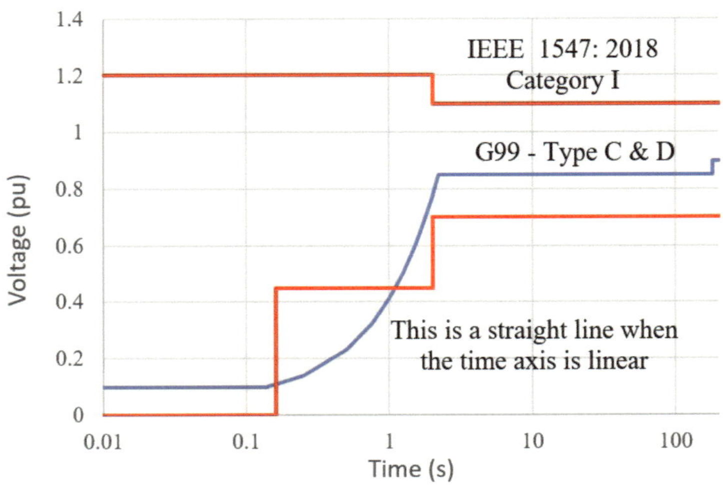

Figure 21.
Fault-ride through requirements.

5.5.1 Fault on utility network: Islanding of the microgrid

A fault at F3 (at 1 s) is cleared by CB3 and CB4 and the microgrid may continue to operate depending on the active and reactive power balance within the microgrid. **Table 7** gives the voltages of different busbar before and after islanding. As can be seen from the table, even after islanding, the voltages of the busbars will be maintained within limits.

Figure 23 shows MS1 terminal voltage and generator frequency when the microgrid is islanded due to a fault at F3. During islanded operation, the total generation within the microgrid is P = 5 MW and Q = 1.5 Mvar whereas the total demand is P = 4.3 MW and Q = 1.5 Mvar. Therefore, the surplus of generation (without considering the losses) after islanding is 700 kW. For MS1 on 5 MVA base, the inertia constant is 2.5 MWs/MVA. From Eq. (7), the rate of change of frequency can be calculated as follows:

$$\frac{df}{dt} = \frac{f}{2H}(P_m - P_e) = \frac{50}{2 \times 2.5} \times \frac{700}{5000} = 1.4\text{Hz/s} \tag{8}$$

From **Figure 23**, it is computed that the rate of change of frequency is 0.036 Hz/s. This is because, when accounting for the losses surplus is much less that 700 kW. Further, when considering the reactive power balance, there is no significant difference to vary the voltage. Therefore, neither under-voltage or over-frequency relays will detect the islanding situation. Typically, the ROCOF relay on a small microgrid is set to operate for 1–3 Hz/s, and with the penetration of micro-sources are increasing it is reduced. For example, in the GB ROCOF setting has recently

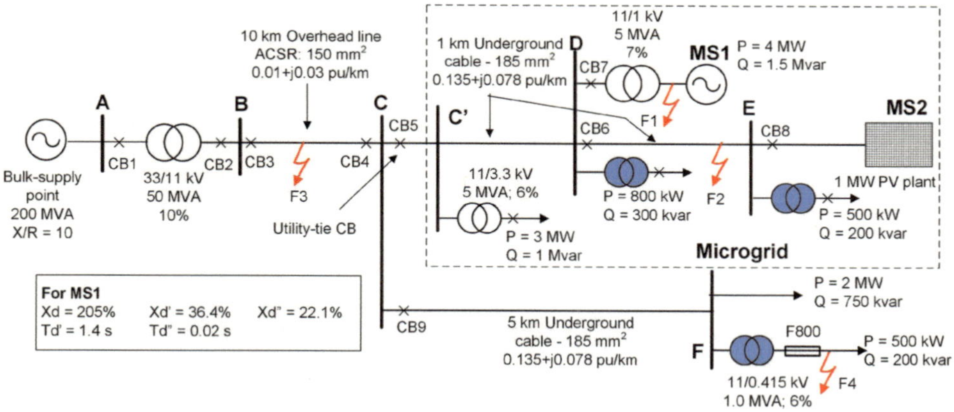

Figure 22.
Example grid-connected microgrid.

Busbar	Voltage (pu) before islanding	Voltage (pu) after islanding
A	1.02	
B	$1.016\angle -0.06^o$	
C	$1.015\angle -0.06^o$	$0.966\angle -3.53^o$ (C′)
D	$1.018\angle -0.01^o$	$0.969\angle -3.49^o$
E	$1.018\angle -0.02^o$	$0.97\angle -3.47^o$

Table 7.
Voltages of different busbars before and after islanding.

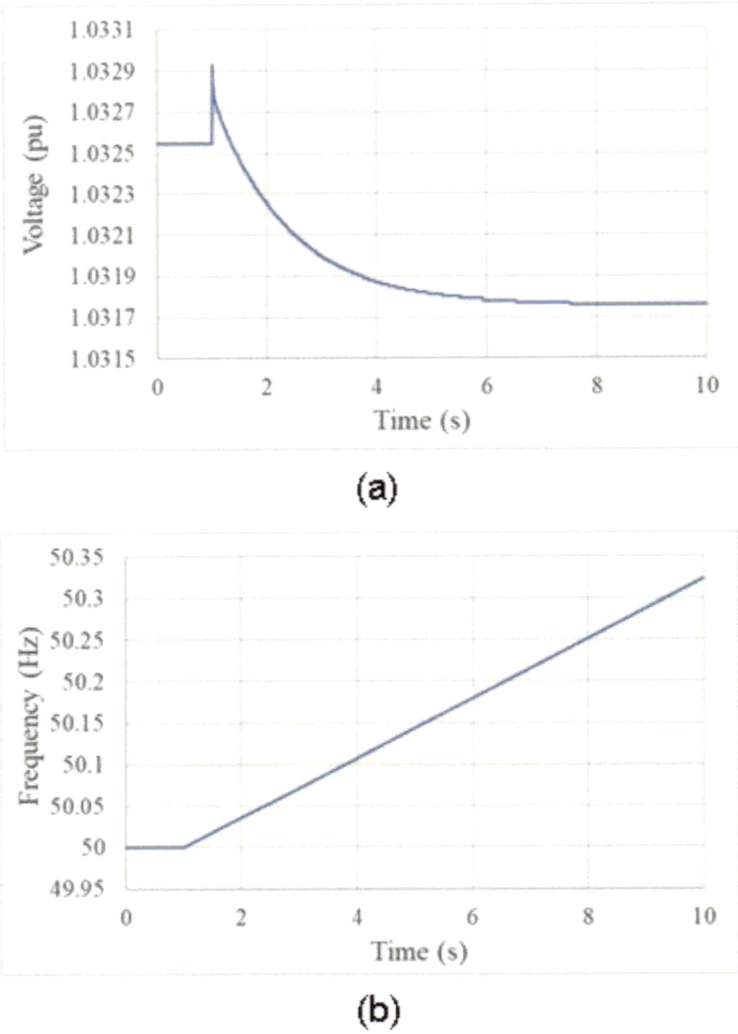

Figure 23.
The voltage and frequency of MS1 terminal. (a) Generator busbar voltage and (b) generator frequency.

changed from 1.25 Hz/s 0.125 Hz/s. With a such setting, the ROCOF relay will not operate. Therefore, the microgrid can continue to operate in islanding mode provided that there is no significant change in the generation and load.

5.5.2 Fault external to microgrid: fault-ride-through of generator in microgrid

Figure 24 shows the voltage at the busbar D for a fault at F4 with grid connection, which is introduced at 1 s and cleared at 2 s. As can be seen, the voltage is well within the ride-through requirements specified in **Figure 21**, and therefore MS1 should ride-through the fault.

5.5.3 Fault internal to microgrid: operation of protection in grid-connected and islanded mode

For a fault at F2 (busbar E), the fault current through CB6 without MS1 is 4.5 kA, and with MS1, it is 7.76 kA. Therefore, the micro-source MS1 aid the protection by reducing the operating time of CB6. With the protection coordination settings for

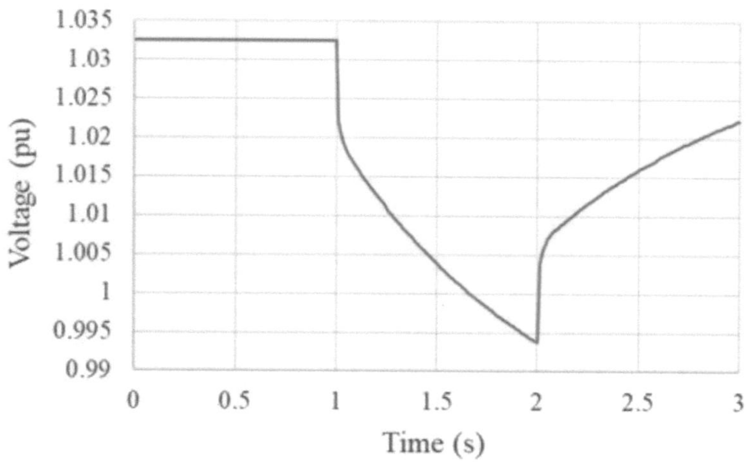

Figure 24.
Voltage at busbar D.

grid connected mode, the protection of CB6 may have mal-operate if the microgrid becomes islanded. In islanded mode, the fault current flowing through CB6 for a fault at F2 reduces to 3.38 kA. The following section explains this issue further.

In grid connected operation, relays at CB9 and CB3 should be coordinated with the fuse F800, which is a 800 A fuse. Then CB5 and CB6 should be coordinated with CB3. These settings are given in **Tables 8** and **9**.

In **Table 8**, full load current in line BC was obtained using load flows carried out in IPSA simulation platform with MS1 and MS2 disconnected. This is the case that introduces the highest current through the line. Similarly, the current through CF was obtained. The possible fault currents at each end of lines (neglecting load currents) was obtained by applying faults at busbars B, C and F. In order to select the primary current rating of the current transformer (CT), and overload current of 125% was considered. The secondary current rating is based on the relay to which the CT is connected and assumed as 5A. Usually, the plug setting (PS) is selected higher than the overload current and less than the fault current and 100% is selected for both relays, i.e. relay at CB3 and CB9. Plug setting multiplier (PSM) is the fault current divided by the plug setting, and it is calculated for each relay for faults at the sending and receiving end of the line.

CB9 should be coordinated with the 800 A fuse, the fuse operating current was obtained for a LV side fault at busbar F. That is for a fault current of 28.83 kA on the LV side of the transformer, from the fuse characteristic, it was found that the operating time is 5 ms. Assuming a grading margin of 0.25 s, the minimum operating time of relay at CB9 is 0.255 s (0.25 + 0.005). Assuming an extremely inverse characteristic [16] for the relay at CB9, the operating time of the relay CB9 was calculated as follows.

$$\text{The operating time of the relay} = 80 \times TMS/\left[PSM^2 - 1\right] \quad (9)$$
$$= 80 \times TMS/\left[28.3^2 - 1\right] = 0.255\ \text{s}$$

where 28.3 is the PSM of the relay at CB9 for a fault at F (see **Table 8**).

Therefore the time multiplier setting (TMS) of the relay = 2.55.

As the maximum possible TMS for a typical overcurrent relay is less than 1.5, the PS of the relay was increased to 130%. Then the PSM = 5650/(1.3 x 200) = 21.7 and the new TMS = 1.49. Select TMS as 1.5 (in many relays, TMS can set only in discrete steps; for this example, it was assumed that the step size is 0.025).

Line	BC		CF	
Line End	B	C	C	F
Full load current (A)	382.3 A (when MS1 and MS2 are not operating)		142.4 A	
Fault current (kA)	8.46	8.07	8.07	5.65
CT ratio[1]	500:5		200:5	
Plug setting (PS)	100% = 500 A		100% = 200 A	
PSM = FC/PS	16.9	16.1	40.4	28.3

[1]*Assuming 125% of overload current.*

Table 8.
Data for selecting the relay for the coordination path F800-CB9-CB3.

Line	CD		DE	
Line End	C	D	D	E
Full load current (A)	195 A (when MF1 and MF2 are operating)		28 A	
Fault current (kA)	8.16	8.09	8.09	7.52
CT ratio[1]	300:5		50:5	
Plug setting (PS)	90% = 270 A		90% = 45 A	
PSM=FC/PS	30.2	29.9	179.8	167.1

[1]*Assuming 125% of overload current.*

Table 9.
Data for selecting the relay for the coordination path CB3-CB5-CB6.

With PS of 130% and TMS of 1.5, the operating time of the relay at CB9 for a fault near busbar C was calculated as follows.

$$\text{PSM} = 8070/(1.3 \text{ x } 200) = 31.0 \tag{10}$$

$$\text{Operating time of the relay} = 80 \times 1.5/\left[31^2 - 1\right] = 0.125 \text{ s} \tag{11}$$

By considering the grading margin of 0.25 s, the operating time of the relay at CB3 for a fault near the busbar C should be 0.375 s. Assuming the relay is very inverse [16], the TMS was calculated as follows.

$$\begin{aligned}\text{The operating time of the relay} &= 13.5 \times TMS/[PSM - 1] \\ &= 13.5 \times TMS/[16.1 - 1] = 0.375 \text{ s}\end{aligned} \tag{12}$$

TMS = 0.419, select 0.425.

Even though the above settings work ok for the given fault, when implementing the protection coordination system in the IPSA simulation platform, the curves of relays at CB9 and CB3 intercept. Therefore, the PS of the relay at CB3 was increased to 150%, and the resultant curves are shown in **Figure 25**.

Table 9 shows the PSM calculation for lines CD and DE for a fault at F2 (busbar E).

With the setting of the relay at CB3 calculated for the previous coordination, the operating time of that relay for a fault at C is 0.58 s. To maintain the correct grading margin, for a fault at the end C, the relay at CB5 should operate at less than 0.33 s, i.e. 0.58–0.25 s. Then assuming that CB5 is a very inverse relay, the TMS was calculated as follows.

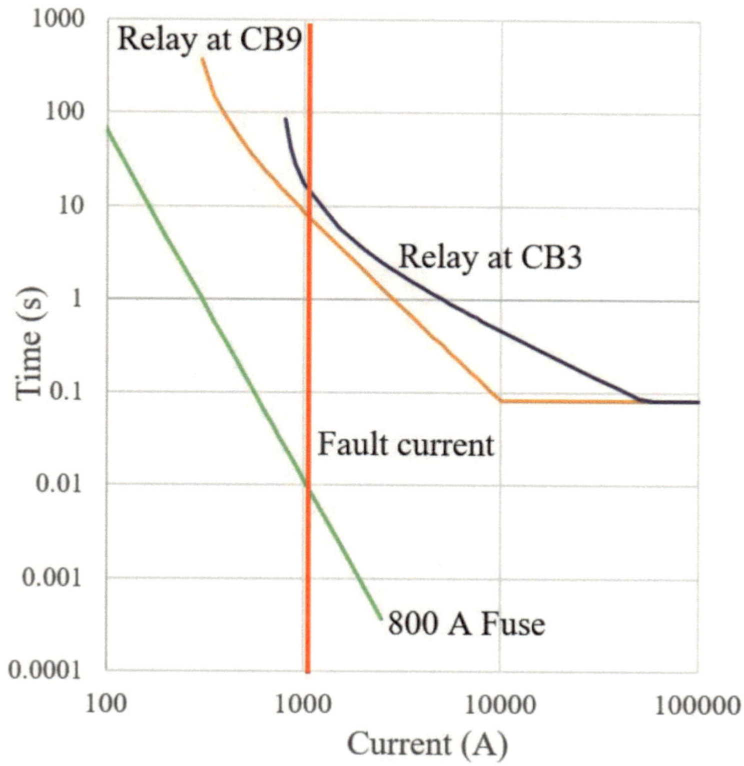

Figure 25.
Relay and fuse characteristic for a fault after 800 a fuse.

$$13.5 \times TMS/[30.2 - 1] < 0.33$$
$$\therefore TMS < 0.71 \tag{13}$$

Therefore, TMS was selected as 0.7.

With the calculated setting of the relay at CB5, the operating time of the relay for a fault at D was calculated as $13.5 \times 0.7/[29.9 - 1] = 0.33$ s. Then the relay at CB6 should operate less than 0.077 s (0.33–0.25 s) for a fault at busbar D. Assuming that CB6 is a very inverse relay, the TMS was calculated as follows.

$$13.5 \times TMS/[179.8 - 1] < 0.077$$
$$\therefore TMS < 1.02 \tag{14}$$

Select TMS as 1.0.s.

Figure 26 shows the coordination of CB3-CB5 and CB6. Also shown the operation of CB6 when MS1 is in operation when MS1 is not in operation, and during an islanded mode for a fault at F2. As can be seen, the operating time of the relay under islanded condition is highest.

6. Earthing requirements

6.1 Effective earthing approaches for transformer connected micro-sources

Micro-sources are operated in parallel with the utility grid and are connected through a transformer. It is important that the proper earthing of this interfacing

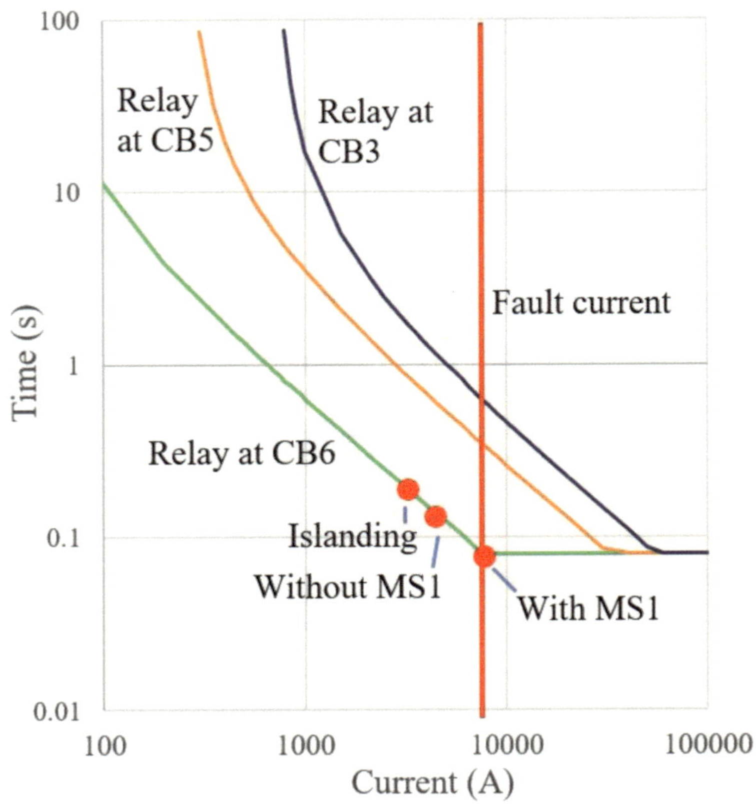

Figure 26.
Relay coordination.

arrangement. **Figure 27** shows possible transformer connection arrangements. It was assumed that the generator is star earthed in all four configurations.

6.1.1 Earthed star: delta connection

This is connection A shown in **Figure 27**. This arrangement provides isolation of the generator from microgrid/utility side ground faults. As discussed in Ref. [17], high zero sequence impedance resulting from this configuration may interfere with the normal flow of fault currents on the microgrid, thus upset protection coordination, nuisance fuse operations, and false operations of upstream protective devices. These problems can be reduced by using a grounding-impedance in the neutral grounding connection of the transformer as per IEEE/ANSI C62.92 [18]. The grounding-impedance should be selected such that the ratio of zero sequence reactance (X0) to positive sequence reactance (X1) less than or equal to 3 and the ratio of zero sequence resistance (R0) to positive sequence reactance less than or equal to 1. This can be easily achieved if the utility side of the interfacing transformer is star-earthed.

6.1.2 Star-star connection

This is connection B shown in **Figure 27**. In order to provide effective grounding of the generator with respect to the utility grid, $X0/X1 \leq 3$ and $R0/X1 \leq 1$ [18]. In this configuration, interferences, especially triplen harmonics, transfer from the grid side to the micro-source side and vice versa.

6.1.3 Delta: earthed star and Delta–Delta connections

These are connections C and D shown in **Figure 27**. If the network on the utility side is ungrounded or delta, then these configurations might create ground overvoltages during faults. To explain this, a part of the microgrid shown in **Figure 28** is used. If the HV side voltage is interrupted, the substation CB opens, and the microgrid may go into the islanded mode. This was a common incident in the Moragahakanda hydropower plant in Sri Lanka. This particular power plant has 4 generators totalling 25 MW and is connected to a 33 kV network. During the first two years of operation, this power station operated as an island with captive loads when the Naula grid, to which the power station is connected, failed. This was later mitigated using a transfer trip arrangement that sends the tripping command to Moragahakanda when the Naula trips.

During the above operating scenario, during a single phase to ground fault the virtual neutral formed in the delta shifts to the original position of phase C as in the phasor diagram shown in **Figure 29**. Then phases-A and -B may experience an

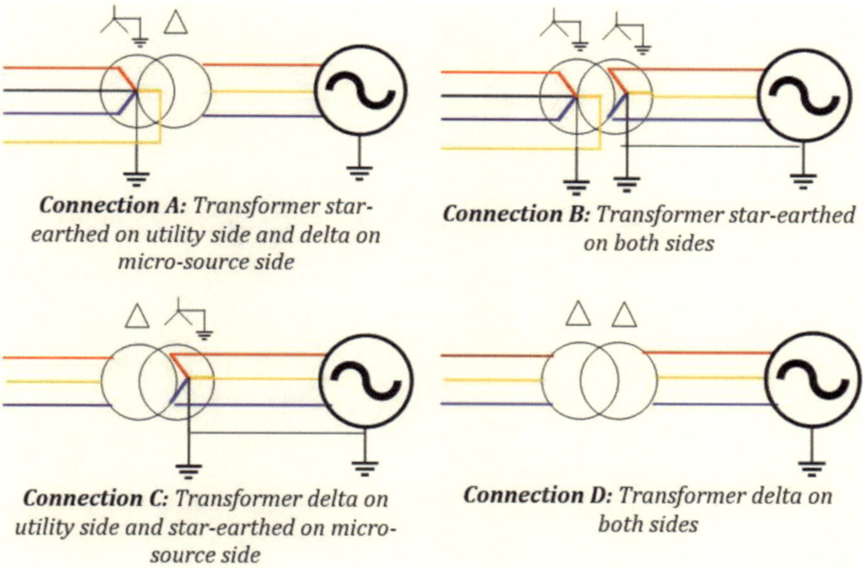

Figure 27.
Possible transformer connection to a micro-source.

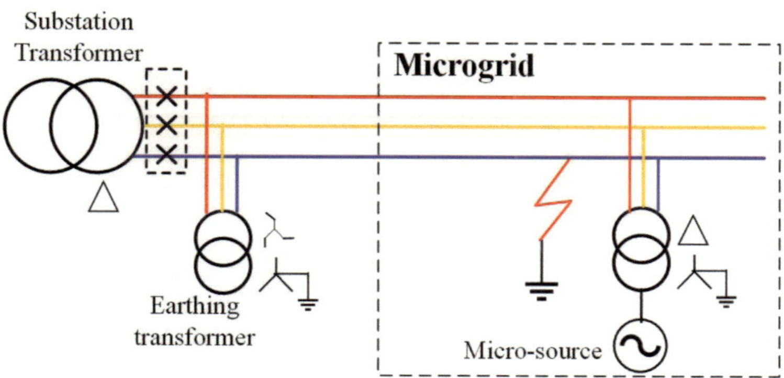

Figure 28.
Single-phase fault on the islanded microgrid.

over-voltage as high as 173% of the original voltage. In practice, this is limited by the faulty impedance and the impedance shown to the fault from the earthing transformer. However, the over-voltages created by such a fault may damage the loads connected to the 33 kV network.

Connection arrangement C can be used with a substation having a Star-earthed, as shown in **Figure 30**. The transfer trip provides fast disconnection of the micro-source, thus preventing any over-voltages appearing on the circuit.

6.2 Effective earthing approaches for VSC connected micro-sources

When paralleled with strong voltage sources (i.e., synchronous generators), VSCs have little impact on neutral grounding, and conventional methodologies of

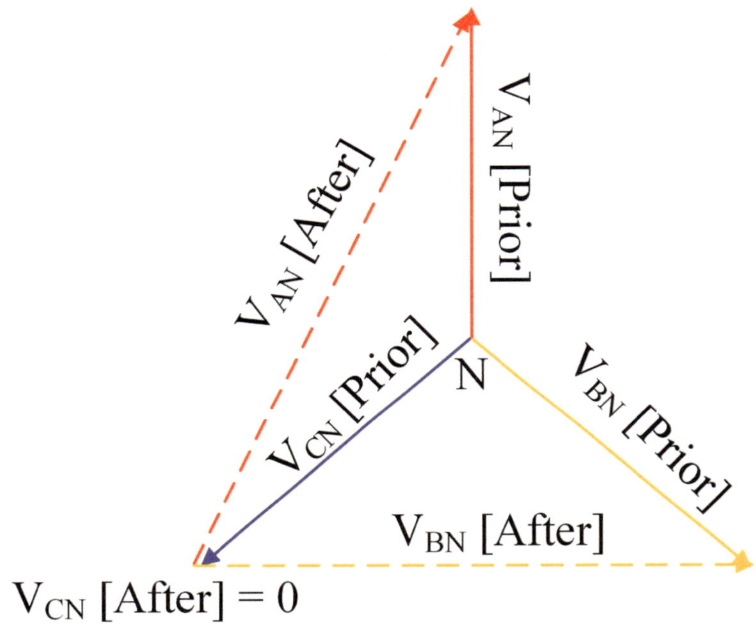

Figure 29.
Phasor diagram for the situation shown in ***Figure 28****.*

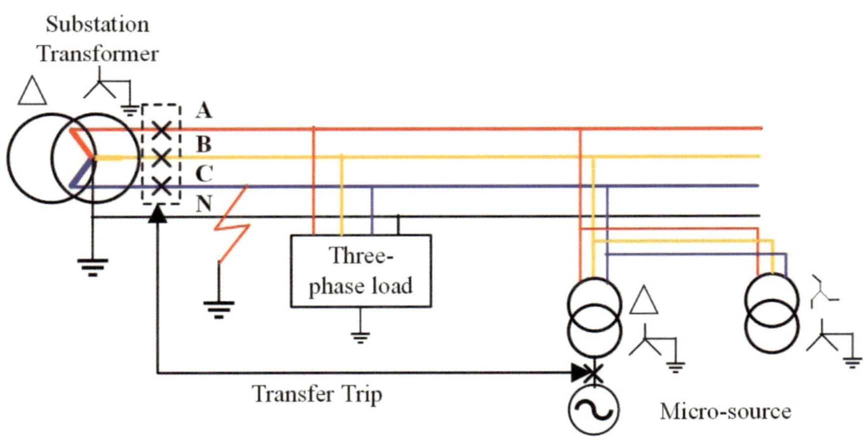

Figure 30.
Typical earthing arrangement for the transformer connection C.

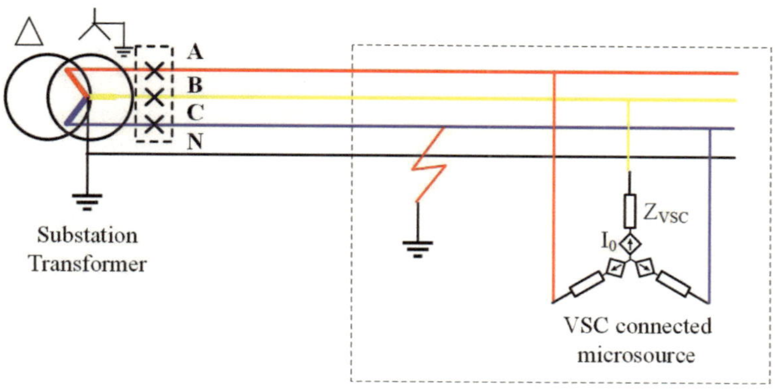

Figure 31.
Representation of a VSC for a single-phase fault on a distribution network.

grounding calculation can be utilised. In most cases, VSCs interconnected with a system providing a strong short-circuit source can be ignored in grounding calculations because of their relatively small short-circuit current contribution compared to that provided by synchronous generators in an interconnected power system. There are certain circumstances where VSCs can become dominant within a sub-system when that sub-system becomes isolated from the main grid. These circumstances involve distribution feeders or systems having sufficient micro-source capacity to support significant energisation of the sub-system when islanded from their normal source.

As most solar inverters are single phase or three-phase three wire, there are no zero sequence currents from them. Therefore, for a transient over-voltage condition that occurs due to single phase fault, it is accurate to assume that the VSC as a positive sequence current source. This is shown in **Figure 31**. Due to the earth fault, if the substation breaker opens and the inverter continues to provide a positive sequence current, then the voltage on health phases is I_0 times the equivalent load impedance between that phase and neutral. A series of tests done on a number of inverters show that this voltage is often greater than 1 pu voltage [19]. When the generation to load ratio is high, a supplementary ground source may be used to reduce the transient over-voltages. IEEE/ANSI C62.92.6 [20] indicates that the supplementary ground sources might interface with the feeder protection coordination, and therefore, case-specific analysis is required.

7. Conclusions

With the formation of generation clusters connected to distribution networks, there is a high possibility of adopting smaller controllable structures such as microgrids operating in parallel to the main utility grid. Even though independent microgrids are in operations since the early ages of electricity generation, the idea of grid connected microgrid immerged more recently.

The protection of independent microgrids is somewhat similar to the protection of special installations. In them, the RCCDs are used to protect against direct and indirect contacts, MCBs or MCCBs are used to protect against overload and short circuit currents, and SPDs are used to protect against over-voltages.

The protection of grid connected microgrids depends on the complexity of the microgrid. Internal faults of micro-sources and their interfacing transformers are protected by a differential protection scheme. Overcurrent protection for faults within the microgrid network and outside is provided by properly coordinated overcurrent protection devices. As micro-sources provide limited fault current voltage operated or voltage restrained overcurrent devices are used. The applicable standards such as IEEE 1547: 2018 and G99 state that under an islanding situation due to a fault outside the microgrid, the microgrid should cease to operate with the times specified in the standards. To provide loss of main protection, under- and over-voltage and under- and over-frequency relays are employed. However, as these relays do not provide adequate protection under LoM, ROCOF relays are used to disconnect the microgrid. For remote faults outside the microgrid, micro-sources should ride-through the fault, and ride-through requirements are specified in the standards.

For correct operation of the microgrid, attention should be paid to the grounding arrangement. The grounding varies with the interfacing arrangement, i.e. transformer connection arrangement or voltage source connection arrangement. In many cases, effective earthing should be maintained as per IEEE/ANSI C62.92.

Acknowledgements

The author would like to acknowledge Mr. Amalendran, Chief Engineer, CEB, Northern Province providing information and pictures of Eluvithive island microgrid. Special thank goes to Eshan Karaunaratne for providing some simulation results and Gayan Abeynayake for checking some calculations.

Author details

Janaka Ekanayake[1,2]

1 University of Peradeniya, Sri Lanka

2 Cardiff University, Cardiff, United Kingdom

*Address all correspondence to: ekanayakej@cardiff.ac.uk

References

[1] Microgrids 1: Engineering, Economics, & Experience, Working Group C6.22, CIGRE publication 635, October 2015, ISBN: 978-2-85873-338-5

[2] Kimbark, E.W., "Power System Stability – Synchronous Machines" (IEEE Press, 1995)

[3] Kundur, P., "Power System Stability and Control" (McGraw-Hill, 1994).

[4] Hindmarsh, J., "Electrical Machines and their Applications" (Pergammon Press, 1970)

[5] Dominik Mamcarz, Paweł Albrechtowicz, Natalia Radwan-Pragłowska and Bartosz Rozegnał, "the analysis of the symmetrical short-circuit currents in backup power supply systems with low-power synchronous generators", Energies 2020, 13, 4474; doi:10.3390/en13174474.

[6] Griffith Shan, M., "Modern AC generator control systems, some plain and painless facts", IEEE Transactions on Industry Applications, 1976, 12(6), pp. 481-491

[7] Tom Key, Gefei Kou, and Michael Jensen, "On Good Behavior: Inverter-Grid Protections for Integrating Distributed Photovoltaics", IEEE PES Magazine, Nov/Dec 2020, pp. 75-85.

[8] IEC62548: 2016: Photovoltaic (PV) arrays – Design requirements, International Electrotechnical Commission, ISBN 978-2-8322-3635-2.

[9] IEC 60364-7-712: 2017: Low voltage electrical installations – Part 7-712: Requirements for special installations or locations – Solar photovoltaic (PV) power supply systems, International Electrotechnical Commission, ISBN 978-2-8322-4006-9.

[10] IEC 61730-2: 2007: Photovoltaic (PV) module safety qualification – Part 2: Requirements for testing, International Electrotechnical Commission, ISBN 978 0 580 67895 0

[11] IEC 61400-1: 2005: Wind turbines – Part 1: Design requirements, International Electrotechnical Commission, ISBN 0 580 47146 2.

[12] WINDSPOT 3.5KW Owner's manual, Available at: 50_030_001_C_MANUAL_windspot_3.5_y_1.5_Manual.pdf (renugen.co.uk); Access on 13/06/2021

[13] Engineering Recommendation G98, Issue 1 Amendment 4 June 2019, "Requirements for the connection of Fully Type Tested Micro-generators (up to and including 16 A per phase) in parallel with public Low Voltage Distribution Networks on or after 27 April 2019", Energy Networks Association.

[14] Engineering Recommendation G99, Issue 1 Amendment 5 November 2019, "Requirements for the connection of generation equipment in parallel with public distribution networks on or after 27 April 2019", Energy Networks Association.

[15] IEEE 1547: 2018, "IEEE standard for interconnection and interoperability of distributed energy resources with associated electric power systems interfaces", IEEE Standards Coordinating Committee 21.

[16] IEC 60255-151: 2009: Measuring relays and protection equipment – Functional requirements for over/under current protection, International Electrotechnical Commission, ISBN 978 0 580 59473 1.

[17] Phil Barker, "Over-voltage Considerations in Applying Distributed Resources on Power Systems", Power Engineering Society Summer Meeting, 2002 IEEE, Volume: 1.

[18] IEEE Std C62.92.1: 2016: IEEE guide for the application of neutral grounding in electrical utility systems—Part I: Introduction, IEEE power and energy society, ISBN 978–1–5044-3735-6.

[19] Nelson, A. Hoke, and S. Chakraborty, J. Chebahtah, T. Wang, and B. Zimmerly, "inverter load rejection over-voltage testing: SolarCity CRADA task 1a final report", Technical Report NREL/TP-5D00–63510, February 2015.

[20] IEEE Std C62.92.1: 2016: IEEE guide for the application of neutral grounding in electrical utility systems—Part VI—Systems supplied by current-regulated sources, IEEE Power and Energy Society, ISBN 978-1-5044-4315-9.

Chapter 2

Power Electronic Converters for Microgrids

Wenlong Ming

Abstract

Power electronic converters are indispensable building blocks of microgrids. They are the enabling technology for many applications of microgrids, e.g., renewable energy integration, transportation electrification, energy storage, and power supplies for computing. In this chapter, the requirements, functions, and operation of power electronic converters are introduced. Then, different topologies of the converters used in microgrids are discussed, including DC/DC converters, single-phase DC/AC converters, three-phase three-wire, and four-wire DC/AC converters. The remaining parts of this chapter focus on how to optimally design and control these converters with the emerging wide-bandgap semiconductors. Correlated tradeoffs of converter efficiency, power density, and cost are analyzed using Artificial Neural Networks to find the optimal design of the converters.

Keywords: power electronics, DC/AC, DC/DC, control, topology, wide-bandgap semiconductor, multi-objective design

1. Introduction

Power electronic converters are essential building blocks in a microgrid, which enable the connection into microgrids of renewable energy resources, energy storage systems, and electric vehicles (EVs), [1–3]. A power electronic converter consists of power semiconductor switches, passive components (inductors, capacitors, transformers, etc.), and a control unit to manage the power conversion and power flow. The main role of power electronic converters is to convert power from one form to another. In addition, power electronic converters can achieve flexible control of active/reactive power fed into the microgrid [4], maximum power point tracking (MPPT) of photovoltaic (PV) cells [5], and wind turbines [6]. The flexible controllability of power electronic converters in microgrids also enables high-level computation and optimization of the microgrid operation and management [7, 8].

Typical power electronic converter-based microgrids are shown in **Figure 1**. This shows a hybrid AC/DC microgrid, which consists of an AC microgrid and a DC microgrid interconnected by an interfacing converter. The microgrid works in grid-connected mode when the utility grid is connected to the AC bus and in stand-alone mode when the utility grid is disconnected.

In an AC microgrid, power electronic converters are used to convert DC power (from PV cells, batteries, EVs, etc.) or variable frequency AC power (from wind turbines) into 50/60 Hz AC power so that the power can be fed into the AC bus and supply loads. In a DC microgrid, power electronic converters are used to convert AC power or DC power with different voltages into DC power with the same

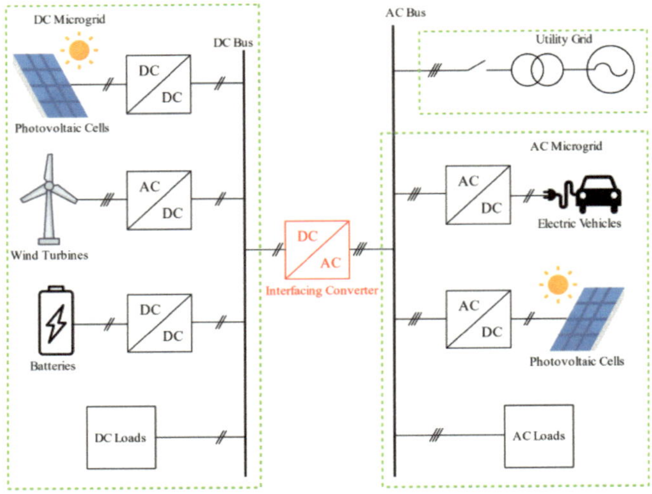

Figure 1.
Power electronic converters in microgrids.

voltage as the DC bus of the DC microgrid. As shown in **Figure 1**, power electronic converters can also be used as the interface between AC and DC microgrids to manage the power flow [9].

Power electronic converters in microgrids use various topologies, according to different applications. Based on the input and output power, power electronic converters can be classified as DC/DC converters and DC/AC converters. DC/DC converters convert the input DC voltages and currents into controlled output DC voltages and currents. DC/AC converters convert DC inputs into controlled AC outputs or vice versa. In microgrids, the DERs can be either DC power or variable AC power.

The control of the power electronic converters ensures that the microgrid functions well in all circumstances [9–11]. In the grid-connected mode when a microgrid is connected to the utility grid, converters in the microgrid operate in grid-feeding mode to provide active and reactive power from distributed generators to the microgrid [12]. In the stand-alone mode when a microgrid is isolated from the utility grid, converters in the microgrid operate in grid-forming or grid-supporting modes to provide AC voltage and frequency support to the microgrid [13]. Besides, converters connected to the energy storage systems and EVs also need to work in battery charging mode to charge the batteries [14]. The different operation modes of power electronic converters require different control algorithms. Converters in the microgrid need to switch between different operations modes according to the status of the microgrid and distributed generators [15, 16].

The design of power electronic converters attempts to meet the system requirements of efficiency, power density, costs, and reliability [17]. However, these factors usually contradict each other. For example, an increase in reliability usually implies higher equipment costs. The power density of a converter can be increased by increasing the switching frequency, which, however, inevitably brings increased switching losses and reduces the overall efficiency. Therefore, tradeoffs must be made between these factors, which requires a multi-objective optimization to find the Pareto-optimal solutions so that the designers can choose the most suitable solution to meet the requirements of a specific application [18].

The rest of this chapter is organized as follows. Firstly, topologies of power electronic converters for microgrids are introduced, including their working principles and applications. Then, an advanced design methodology of power electronic converters based on multi-objective optimization considering the cost, efficiency, and power density is presented. Finally, the control algorithms of power electronic converters for different operation modes in the microgrid are summarized.

2. Converter topologies

2.1 DC/DC converters

DC/DC converters are used to create either a higher or lower constant DC voltage at the output without being affected by possible fluctuations of the DC input voltage or of the load current. These converters are essential for the operation of devices in various industry areas such as renewable energies, automotive, battery chargers and aerospace.

They use semiconductors that behave like switches that are opened and closed to convert an input voltage level into a different one, applying a series of control signals. Generally, the way the voltage is converted is by forcing energy to be stored in an inductor or capacitor When the polarity is reversed, the energy stored is discharged creating a voltage at the converter output.

Among the different control methods, the most popular one is pulse width modulation (PWM), which regulates the output voltage through adjusting the on and off times of the semiconductors. A Duty Ratio D is defined as the ratio between the on and off time of the semiconductors in a switching period. The output voltage is regulated by controlling the value of D, which is between 0 and 1 [19].

2.1.1 Buck converter

The buck converter is shown in **Figure 2**. creates an output voltage that is lower than the input voltage V_{in}. The average output voltage V_o is a function of the duty ratio according to the expression:

$$V_o = D\,V_{in} \tag{1}$$

The voltage V_o is controlled by varying the duty ratio D of the switch S. This is a linear relationship for ideal conditions. When the switch is on, the diode is reverse biased and so the input current flows through the inductor L. When the switch is off, the diode conducts so the energy stored in the inductor passes through the diode supplying part of this stored energy to the load. The capacitor at the output is used to keep the average value of the output voltage constant [20].

2.1.2 Boost converter

The boost converter in **Figure 3** shows the circuit topology, where the main function is to obtain a higher DC voltage at the output than its input DC voltage. It contains at least one semiconductor switch and elements to store energy.

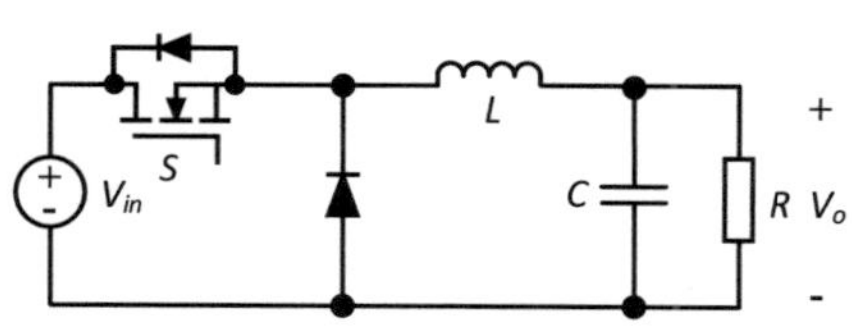

Figure 2.
Buck converter.

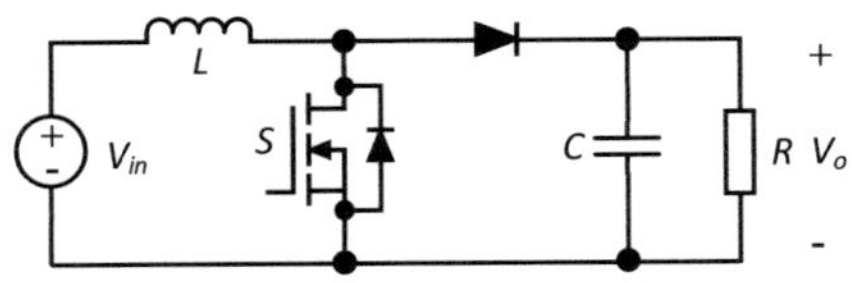

Figure 3.
Boost converter.

When the switch S is closed, the inductor stores energy from the source, at the same time the load R is fed by the capacitor (C). When the switch is open, the only path of the current is through the diode, and current flows to the capacitor and the load [21].

$$V_o = \frac{V_{in}}{1 - D} \tag{2}$$

Since the output voltage is inversely proportional to $(1 - D)$, a higher duty ratio gives a higher output voltage.

2.1.3 Buck-boost converter

A buck-boost converter is used if the desired output voltage may be higher or lower than the input DC voltage.

Figure 4 shows a buck-boost converter, which is a cascaded connection between a buck converter and a boost converter. In steady-state, the ratio between the input and output voltages is the product of the ratios of both converters as:

$$\frac{V_o}{V_{in}} = \frac{D}{1 - D} \tag{3}$$

When the switch is closed, the input provides power to the inductance and the diode is reverse biased. When the switch is open, the energy stored in the inductance is transferred to the output while the input does not provide power to the load. As in the previous converters, in steady-state the capacity of the output capacitor is large enough so that the output voltage Vo is constant [22]. The output voltage is reversed inside.

2.2 DC/AC single-phase converters

A DC/AC converter also known as an inverter is capable of transforming DC into AC. A single-phase inverter is shown in **Figure 5**. It is often used, for small loads and power supplies.

Single-phase inverters are fundamental parts of solar PV generating systems that convert from DC into AC, allowing the power generated to be supplied to a network with fixed voltage and frequency [23]. Another application is in battery storage

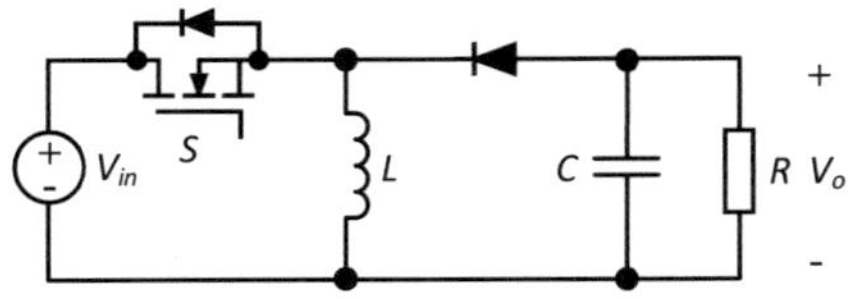

Figure 4.
Buck-boost converter.

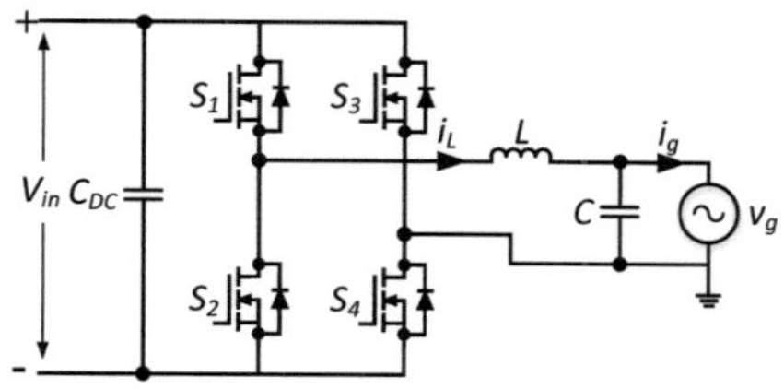

Figure 5.
Singe-phase DC/AC inverter.

systems, which are able of accumulating energy for later use when there is not enough solar resource to cover the load demand, using a single-phase inverter to obtain the voltage level and the frequency necessary to be connected to the main grid [24, 25].

Single-phase inverters when connected to the grid under unbalanced conditions, can propagate second-order harmonics from AC to DC side. The power for the grid side P_g is defined as:

$$P_g = V_g I_g = VI\cos^2\omega t = \frac{VI\,(1 + \cos 2\omega t)}{2} = \frac{VI}{2} + \frac{VI}{2}\cos 2\omega t \tag{4}$$

V_g and I_g represent the grid voltage and current respectively, meanwhile, the angular grid frequency f is defined as $\omega = 2\pi f$. As second-harmonics are present in the DC side, large DC capacitance is required to support second-order ripples [26].

2.3 Three-phase DC-AC converters

The principle for operating three-phase DC-AC converters is similar to that of a single-phase converter. The essential difference is that instead of having a single modulated signal, there are three sinusoidal modulated signals out of phase 120° between them. These signals are compared with a triangular signal for the control of the switches at three phases. The voltage signals obtained in the phases with respect to the neutral are identical to those of the single-phase modulation, but instead of having a single signal, there are three signals out of phase by 120°. Therefore, the same theory can be applied for single-phase converters.

A conventional two-level three-phase inverter consists of six switches generating an output voltage with two values with respect to the negative terminal of the input capacitor as shown in **Figure 6** [27].

The development of 3-level or multilevel technology has been influenced by the development of new wide-bandgap materials for the construction of semiconductors as IGBT and MOSFET, allowing high voltage and current operation and higher switching frequency obtaining lower harmonic distortion at the output [28].

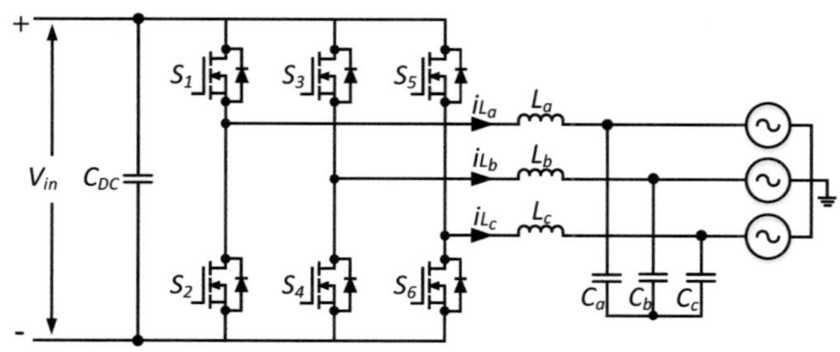

Figure 6.
Three-phase two-level DC/AC inverter.

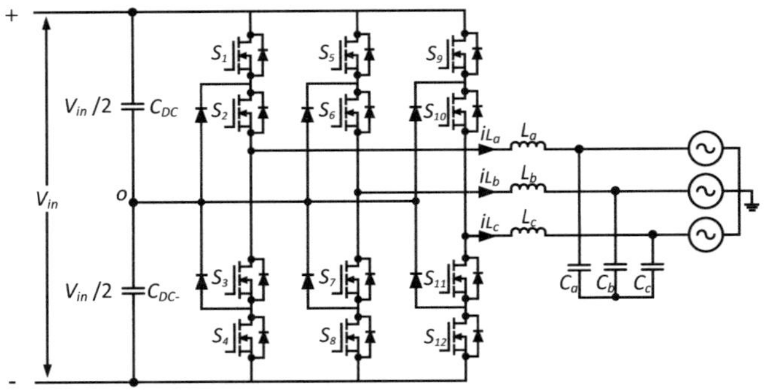

Figure 7.
Three-phase three-level DC/AC neutral-point-clamped inverter.

A popular three-level inverter is the Neutral Point Clamped Converter. This inverter can be considered as the origin of the recent multi-level inverter and has been widely studied and applied. The topology is shown in **Figure 7**.

In this type of inverter, each leg consists of four switching devices clamped by two diodes. The DC bus voltage is divided by two capacitors connected with the clamped diodes, granting the neutral point (o). The output voltage in each phase can take three levels, $V_{in}/2$, 0, and $V_{in}/2$.

The diodes connected to the midpoint are the elements that set the blocking voltages of the switches to a fraction of the DC bus voltage; therefore, they are the key elements of this topology. This topology can be extended to more levels but the number of switching devices used is greatly increased [29].

2.4 Three-phase four-wire converters

Three-phase four-wire converters consist of three half-bridges that distribute phases A, B, and C, and an additional leg allows the path of the neutral currents.

One of the most attractive applications of these inverters is electric vehicle EV chargers, allowing both charging the EV in the grid to vehicle G2V mode and vehicle to grid mode V2G. The V2G mode can generate an autonomous network by obtaining energy from the EV battery to the grid when there are high energy demands or blackouts, being EV chargers susceptible to power quality problems when there is an imbalance due to loads in the grid [30].

Ideally, the operation of a distribution system is that all the loads that are connected are of the same magnitude which results in the balanced operation of the system. However, any practical system is always operating in unbalanced conditions, leading to overheating, losses and malfunction in transformers, protection systems and cables connected to the network.

Three common topologies being used in microgrids for four-wire systems are Split DC-Link, four-leg three-phase inverter, and independently-controlled neutral leg.

2.4.1 Three-phase four-wire inverter with Split DC-link capacitor

This basic typology consists of only six switches S_1-S_6 as seen in **Figure 8**. containing a neutral point being clamped to the DC bus inverter, resulting in two voltage levels through the DC-link capacitors.

Zero-sequence current causes voltage ripple at the midpoint of the inverter, therefore large capacitors are required to reduce the voltage ripples. To share the

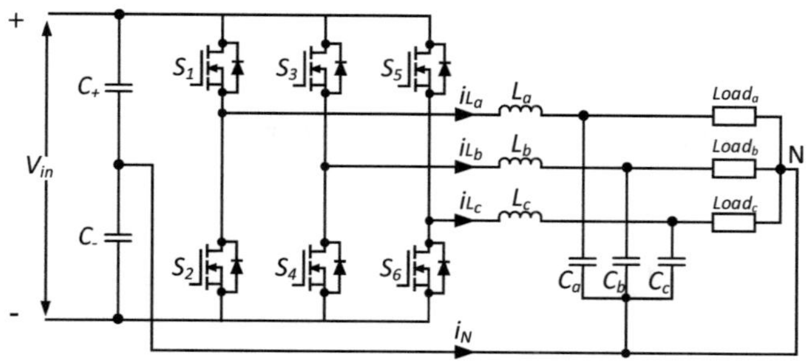

Figure 8.
Three-phase four-wire DC/AC inverter with split DC-link capacitors.

input DC voltage evenly, the voltages of the two divided DC link capacitors must be controlled [31].

One of the disadvantages of this typology is that large neutral currents, are reflected in voltage ripple in the capacitors.

2.4.2 Three-phase four-wire DC/AC inverter with neutral leg

Figure 9 shows the configuration of this inverter, where switches S_1-S_6 feed the ABC phases, while switches S_{N1} and S_{N2} provide the neutral current line. An advantage of this converter is that the capacitor does not need to be bulky to reduce the second-order ripples in the DC-link, but its control strategy is more complex than the previous converter. Additionally, there is a 15% gain in AC output voltage with respect to the DC voltage. As the additional leg cannot be controlled independently, the control complexity of maintaining balanced voltages on the AC lines as well as maintaining the neutral point can be experienced great stress on the DC terminals causing electromagnetic interference (EMI) [32].

2.4.3 Three-phase four-wire DC/AC inverter with an independently-controlled neutral leg

Similar to the three-phase neutral leg inverter, the topology in **Figure 10** is configured with eight switches in total, being S_1-S_6 for ABC phases and S_{N1}-S_{N2} for a neutral current path to the mid-point between the two DC link capacitors. A large amount of capacitance is not required as there are no high current levels through the DC link capacitors.

This topology is highly preferred for use in unbalanced load conditions since the current flowing through the inductor allows the DC link to remain neutral.

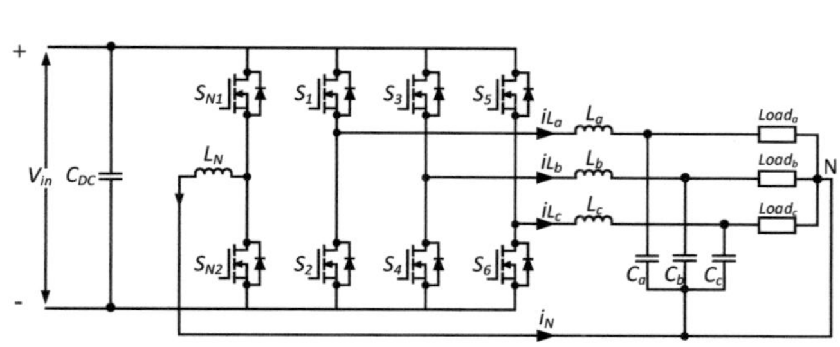

Figure 9.
Three-phase four-wire DC/AC inverter with the neutral leg.

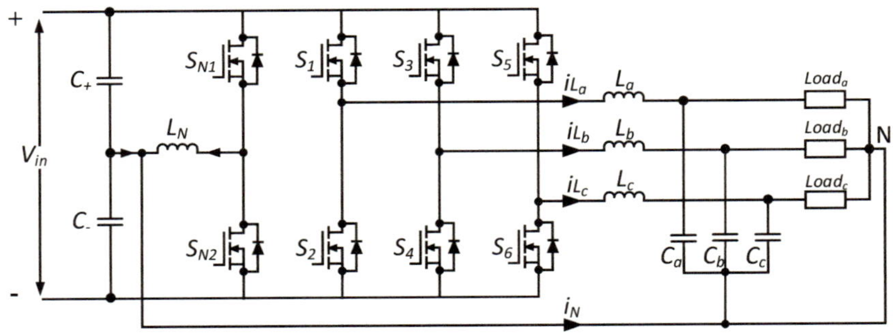

Figure 10.
Three-phase four-wire DC/AC inverter with an independently-controlled neutral leg.

Its control strategy also allows the phases to be controlled independently as well as the neutral leg, avoiding stress on the DC link and other interferences. Another aspect is that by not presenting neutral current through the DC link capacitors; small capacitors can be used and power density will be increased [32].

3. Design of wide-bandgap semiconductors based power electronics converters

3.1 Benefits of wide-bandgap semiconductors

Wide-bandgap (WBG) semiconductors such as silicon carbide (SiC) and gallium nitride (GaN) have superior material properties for power electronics compared to conventional Silicon (Si) as shown in **Figure 11**. The semiconductor properties have been characterized by the electric field, energy gap, electron velocity, melting point, and thermal conductivity.

Power switches based on WBG semiconductors have better switching and conduction performance over a wide range of temperatures in comparison to Si-based devices. For instance, they have a faster switching speed, lower switching losses, higher breakdown voltages, and higher operating temperatures. Therefore, WBG devices are considered promising solutions for high-efficient power electronics designs. These properties enabled the devices to achieve better performance for

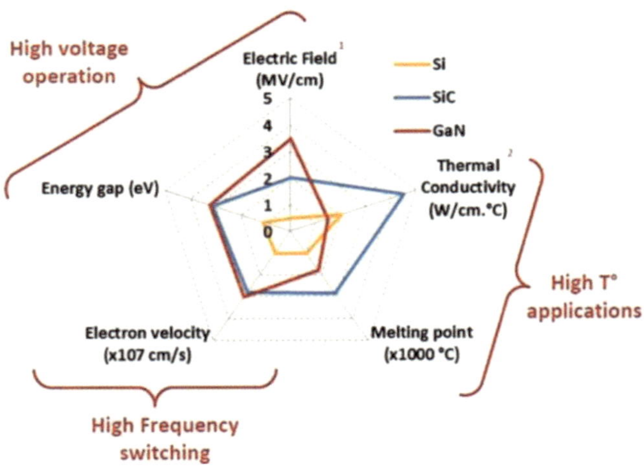

Figure 11.
The properties (and implications for operation) of Si, SiC, and GaN.

applications that require high voltage operation, high temperature, and high switching frequency.

3.2 Design challenges of WBG power converters

A well-designed power electronic converter should have high efficiency, small volume, and lightweight, low cost, and low failure rate. However, the challenge is the trade-off between these performance measures. For example, if a design is focused only to achieve higher efficiency, which impacts the power density, reliability, and cost. The performance parameters must be simultaneously optimized, such as maximizing efficiency while reducing volume, reliability, and cost as much as possible. Typical technical performance measures are efficiency, volume, cost, and reliability as depicted in **Figure 12**. These measures are mainly determined by the design of the converter including selections of topology, modulation scheme, components, and layout. By carrying out the careful design of the converter, the converter will have high power density, as well as reduced volume, weight, and cost.

3.3 Multi-objective design of WBG power converters

Multi-objective design is used to optimize the efficiency and volume of WBG power converters. To achieve this, Pareto fronts are determined. A Pareto front is the absolute limit of performance trade-off for a given set of converter specifications (topologies, control schemes, and components). Detailed mathematical modeling of the converter components is required for determining the Pareto fronts.

3.3.1 Modeling of power loss and volume of the inverter

In a single-phase inverter, the DC-link current is composed of both DC components and second-order ripple current. The second-order ripple current is occurred due to the unbalanced nature of single-phase systems. The conventional DC/AC inverter (see **Figure 5**) requires additional active or passive components to reduce the magnitudes of second-order ripple current. Instead, differential inverters can reduce the ripple current without adding extra components. As a result, Differential inverters are popular solutions for applications where reduction of second-order ripple current is critical and hence the differential inverters are modeled in this study.

Figure 11 shows the GaN switch-based differential inverter topology used to study the multi-objective design. The differential DC/AC inverter topology is

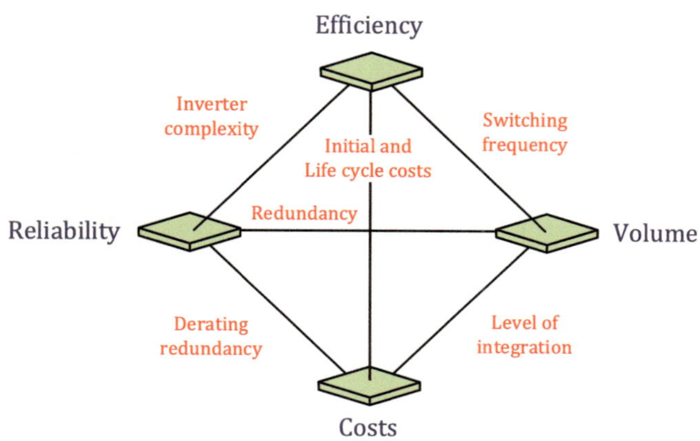

Figure 12.
The trade-off between design and performance parameters.

developed using two bidirectional DC/DC buck converters. In this study, the objectives are to optimize the efficiency and power density of a DC/AC buck inverter. To start with, the efficiency and power density of four major components within an inverter are modeled including the power GaN FETs, inductors, capacitors, and heat sinks. Their efficiency and power density are further determined by the variables including switching frequency $\boldsymbol{f_{sw}}$, the inductor ripple $\boldsymbol{\Delta i_L}$, the switch area $\boldsymbol{A_{sw}}$, and the junction temperature $\boldsymbol{\Delta T_j}$. Therefore, it is of great importance to model the efficiency and power density based on these variables for each component.

3.3.2 Power GaN FETs

The power loss models of the GaN FETs are based on the on-state resistance $\boldsymbol{R_{DS,on}}$, the output capacitance $\boldsymbol{C_{oss}}$ and the thermal junction-to-case resistance $\boldsymbol{R_{\theta JC}}$ of the switches. The output capacitance $\boldsymbol{C_{oss}}$ represents the parasitic capacitance of the power GaN FETs. The value of $\boldsymbol{C_{oss}}$ is provided in the datasheet of GaN FETs. When the energy is stored into the output capacitor, current discharges through the body diode causes power loss [33]. These variables are scaled by their reference values with respect to the area of the switch. The switching loss of the inverter is the sum of the turn-on and turn-off loss of all the switches [33]. The switching loss of the higher side switches $\boldsymbol{P_{S_H,sw}(H = 1, 3)}$ (see **Figure 13**) is obtained as,

$$P_{S_H,sw} = \frac{V_{in} f_{sw}}{2} \left\{ \left(I_{out} sin\,(\omega t) + i_{comp} - \frac{\Delta i_{L_a}}{2} \right) (t_{CR} + t_{VF}) + \left(I_{out} sin\,(\omega t) + i_{comp} + \frac{\Delta i_{L_a}}{2} \right) (t_{VR} + t_{CF}) \right\} \tag{5}$$

Where, i_{comp} is the second-order current component and Δi_{L_a} is the inductor current ripple. t_{CR} and t_{CF} are the rise and fall times for the current in the switch. t_{VR} and t_{VF} are the rise and fall times for the voltage in the switch.

The switching loss of the lower side switch $\boldsymbol{P_{S_L,sw}(L = 2, 4)}$ can be derived as:

$$P_{S_L,sw} = \frac{V_{SD} f_{sw}}{2} \left\{ \left(I_{out} sin\,(\omega t) + i_{comp} + \frac{\Delta i_{L_a}}{2} \right) (t_{CR} + t_{VF}) + \left(I_{out} sin\,(\omega t) + i_{comp} - \frac{\Delta i_{L_a}}{2} \right) (t_{VR} + t_{CF}) \right\} \tag{6}$$

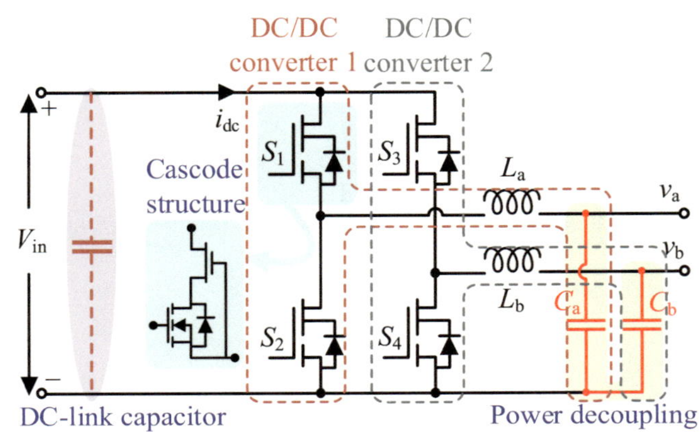

Figure 13.
Structure of a differential DC/AC buck inverter.

From (5) and (6), the total switching losses $\boldsymbol{P_{tot,sw}}$ of the inverter are calculated as the sum of $\boldsymbol{P_{S_H,sw}}$ and $\boldsymbol{P_{S_L,sw}}$. The switching losses of higher side switches depend on the input voltage $\boldsymbol{V}_{\text{in}}$, and lower side switches depend on the diode voltage $\boldsymbol{V}_{\text{SD}}$. Hence, the lower side switches produced lesser switching losses compared to the higher side switches as V_{SD} is much smaller than V_{in}.

The conduction loss depends on the RMS current flowing through the switch $\boldsymbol{I}_{\textbf{RMS,sw}}$, the on-state resistance $\boldsymbol{R}_{\textbf{DS,on}}$ and the change in junction temperature $\boldsymbol{\Delta T_j}$. It will be varied according to the duty cycle of the switches $\boldsymbol{S_1} - \boldsymbol{S_4}$. After applying the mathematical simplifications, the total conduction loss $\boldsymbol{P}_{\textbf{tot,cond}}$ can be written as:

$$\begin{aligned} P_{\text{tot,cond}} = \left(\frac{R^*_{\text{DS,on}} A^*_{\text{sw}}}{A_{\text{sw}}}\right)(1+\Delta T_{\text{j}})\Bigg\{\left(I^2_{\text{out}}\, sin^2(\omega t) + i^2_{\text{comp}} + \frac{\Delta i^2_{\text{L}_\text{a}}}{12}\right) \\ + \left(I^2_{\text{out}}\, sin^2(\omega t+\pi) + i^2_{\text{comp}} + \frac{\Delta i^2_{\text{L}_\text{b}}}{12}\right)\Bigg\} \end{aligned} \tag{7}$$

The power losses of the output capacitance C_{oss}, depend on the input voltage and the switching frequency as:

$$P_{\text{tot,C}_{\text{oss}}} = 2\left(\frac{C^*_{\text{oss}} A_{\text{sw}}}{A^*_{\text{sw}}}\right) V^2_{\text{in}} f_{\text{sw}} \tag{8}$$

The reverse recovery loss of the lower side switches is not negligible for cascode devices. The total reverse recovery loss $P_{tot,rr}$ is calculated as:

$$P_{\text{tot,rr}} = 2\left(\frac{Q^*_{\text{rr}} A_{\text{sw}}}{A^*_{\text{sw}}}\right) V_{\text{in}} f_{\text{sw}} \tag{9}$$

The gate losses depend on the switching frequency, the gate-source voltage V_{GS} and the gate charge Q_g. The total gate loss of four switches $P_{tot,g}$ is calculated as:

$$P_{\text{tot,g}} = 4\left(\frac{Q^*_{\text{g}} A_{\text{sw}}}{A^*_{\text{sw}}}\right) V_{\text{GS}} f_{\text{sw}} \tag{10}$$

In cascode GaN FETs, the current flowing through the body diodes of the lower side switches incur the conduction loss during the reverse recovery time t_{rr} [32]. The total power loss $P_{tot,bd}$ of the body diodes can be written as,

$$P_{\text{tot,bd}} = 2V_{\text{SD}} f_{\text{sw}} t_{\text{rr}} \left(I_{\text{out}}(sin\,(\omega t) + sin\,(\omega t+\pi)) + 2i_{\text{comp}}\right) \tag{11}$$

The volume of the switches can be calculated as:

$$vol_{\text{sw}} = 4h_{\text{sw}} A_{\text{sw}} \tag{12}$$

Where, h_{sw} is the height of the switch package.

3.3.3 Output inductors

The inductor power loss consists of the core loss and the AC and DC resistance loss which can be expressed as [34].

$$P_{ind} = \left(a_{L1} f_{sw}^{\alpha} \Delta i_{L}^{\beta}\right)_{core\_loss} + \left(a_{L2} f_{sw} \Delta i_{L}^{\gamma}\right)_{AC\_loss} + \left(a_{L3} I_{out}^{2} \Delta i_{L}^{\lambda}\right)_{DC\_loss} \quad (13)$$

Where, a_{L1}, α, and β are the Steinmetz coefficients; a_{L2} and a_{L3} are the constants which are used to approximate the values of DC winding resistance; γ and λ are the real values used to reduce the non-linearity. The approximated inductor volume is calculated as:

$$vol_{ind} = a_{L4} L I_{peak}^{2} + a_{L5} L I_{peak} + a_{L6} I_{peak} \quad (14)$$

Where, a_{L4}, a_{L5}, and a_{L6} are the polynomial coefficients of the inductor which must be a positive value. L is the inductor value. I_{peak} is the peak current of the inductors.

3.3.4 Output capacitors

The power loss of the capacitor is calculated as:

$$P_{cap} = \frac{I_{RMS,C}^{2} \tan\delta}{2\pi f_{2\omega} C} \quad (15)$$

Where, $I_{RMS,C}$ is the RMS current flow through the capacitor, $\tan\delta$ is the loss factor, $f_{2\omega}$ is the frequency of second-order ripple power and C is the value of the capacitance. The total volume of the capacitors vol_{cap} is calculated as:

$$vol_{cap} = a_{C1} C V_{C}^{2} + a_{C2} C V_{C} + a_{C3} V_{C} \quad (16)$$

Where, a_{C1}, a_{C2} and a_{C3} are the polynomial coefficients of the capacitors which must be positive values. The voltage V_C is the voltage across the capacitor.

3.3.5 Heat sinks

The volume of the heat sink is calculated as:

$$vol_{heat\ sink} = \frac{V_{\theta SA}}{P_D}\left(\Delta T_j - P_D\left(\frac{R_{\theta JC}^{*} A_{sw}^{*}}{A_{sw}} + R_{\theta CS}\right)\right) \quad (17)$$

Where, $V_{\theta SA}$ is the volumetric resistance, P_D is the power dissipated by the GaN FETs, ΔT_j is the temperature difference between the junction and the ambient, $R_{\theta JC}^{*}$ is the thermal resistance from junction to the case of the semiconductor, and $R_{\theta CS}$ is the thermal resistance from the case to the mounting surface of the semiconductor.

3.3.6 Formulation of the multi-objective model

To formulate the multi-objective model, the total power loss $P_{tot,loss}$ and volume vol_{tot} are calculated as:

$$P_{tot,loss} = P_{tot,sw} + P_{tot,cond} + P_{tot,C_{oss}} + P_{tot,rr} + P_{tot,g} + P_{tot,bd} + P_{ind} + P_{cap} \quad (18)$$

$$vol_{tot} = vol_{sw} + vol_{ind} + vol_{cap} + vol_{heat\ sink} \quad (19)$$

Using (18) and (19), the objective function and inequality constraints can be obtained as:

$$\begin{aligned} &\text{minimize } f(P_{tot,loss}, vol_{tot}) \\ &\text{subject to } f_{sw,\,min} \le f_{sw} \le f_{sw,\,max} \\ &A_{sw,\,min} \le A_{sw} \le A_{sw,\,max} \\ &\Delta i_{L,\,min} \le \Delta i_L \le \Delta i_{L,\,max} \\ &\Delta T_{j,\,min} \le \Delta T_j \le \Delta T_{j,\,max} \end{aligned} \tag{20}$$

From (20), the optimal value of the power loss and volume of the inverters is determined following the iterative process shown in **Figure 14**. The optimized efficiency and power density of the design are then calculated. The outcome of the multi-objective design is the Pareto-front showing the optimized efficiency and power density of the designed inverter.

3.4 Performance evaluation

The multi-objective design approach was implemented in MATLAB/Simulink and examined a 1 kW GaN-based inverter. The performance of the inverter was examined in terms of efficiency and power density. The minimum and maximum values of the design variables used for the multi-objective design are given in **Table 1**. The values of design variables are selected from industrial design standards.

A 900 V GaN FET manufactured by Transphorm was used. A P11T60 series of high current toroid type fixed inductors designed by MPS Industries were used. A MKP1848C series of polypropylene film capacitors from Vishay BC Components

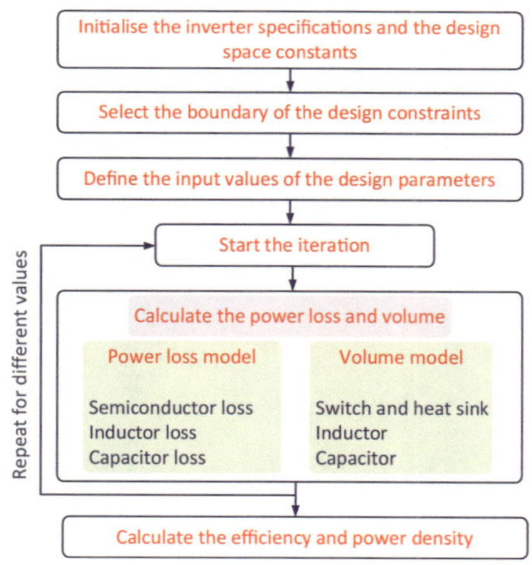

Figure 14.
Flow chart of multi-objective design.

Design variable	Min. value	Max. value
Switching frequency f_{sw}	30 kHz	200 kHz
Current ripple Δi_L	0.1 $I_{out,max}$	0.45 $I_{out,max}$
Switch area A_{sw}	0.94 A^{*}_{sw}	1.07 A^{*}_{sw}
Change in temperature ΔT_j	1 °C	25°C

Table 1.
Design constraints of the inverter.

was used. The values of the maximum output current $I_{out,max,}$ and reference switching area A^{*}_{sw} are 6.15 A and 45.6 mm^2. The values of the inductor are L = 390 μH, the capacitor is C = 48 μF and the switching frequency is f_{sw} = 100 kHz. These are selected according to the outcome of the multi-objective design approach.

3.4.1 Performance evaluation

The Pareto-front performance of efficiency and power density of the power inverter is generated by the multi-objective design and is given as the green curve in **Figure 15**.

One design based on the Pareto-front performance was chosen to validate the method. The selection of efficiency and power density was made based on different applications. For instance, the selected efficiency and power densities are 98.4% and 4.6 kW/dm3 which are typical for PV inverters. For the corresponding design, the power loss and volume are obtained as 15.93 W and 218.32 cm^3. The breakdown of power loss and volume of each component is given in **Figure 16**. With the total power loss, switches contribute 52%, inductors contribute 33% and capacitors contribute 15%. Likewise, with the total volume, heat sinks and switches contribute 34%, inductors contribute 34%, and capacitors contribute 32%.

3.4.2 Experimental verification

A prototype of the inverter was built using the components sized using the multi-objective design. The prototype is shown in **Figure 17(a)**. The prototype was

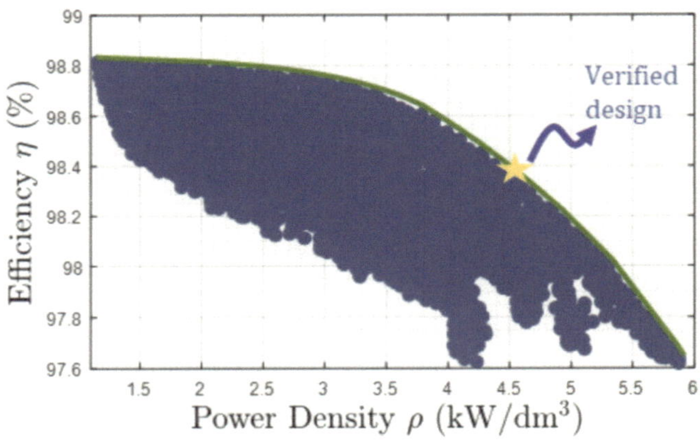

Figure 15.
Efficiency vs. power density.

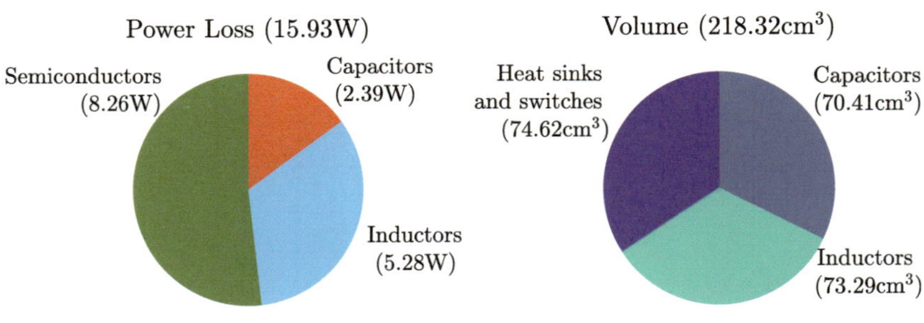

Figure 16.
Power loss and volume.

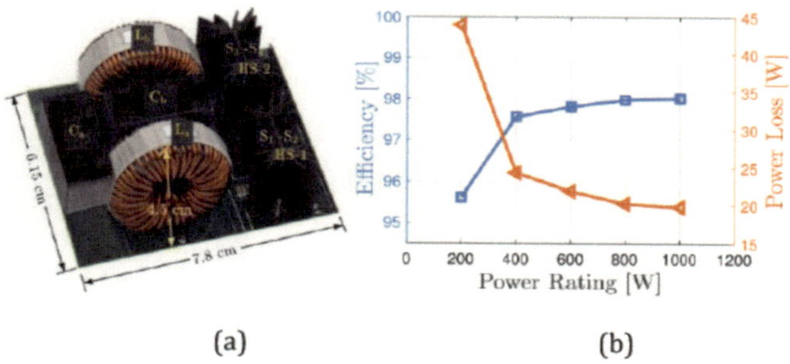

Figure 17.
(a) Hardware prototype, and (b) efficiency and power loss.

operated at different output power levels to obtain the efficiency and power loss and the result is given in **Figure 17(b)**. The efficiency was measured by a Yokogawa WT1806E precision power analyzer. It was observed that the maximum efficiency of the prototype is 98.02%. The power density was 4.54 kW/dm^3 from the volume of the inverter which is given in **Figure 17(a)**. Therefore, the efficiency and power density match the results obtained by the proposed design approach in Section 3.4.1. The errors of efficiency and power density obtained from both are only 0.38% and 0.06 kW/dm^3 respectively.

4. Converter control

For the converter system, an important question is how to design a good controller for the system so that the system can run stably while meeting the required performance indicators. In this section, typical approaches to the control of both DC/DC and DC/AC power electronic converters used in microgrids are presented.

The control of converters usually has a hierarchical control structure (see **Figure 18**). The switching-level, converter-level, and application-level control are introduced in this section. For the switching-level control, typical pulse-width modulation (PWM) methods are introduced. For the converter-level control, the

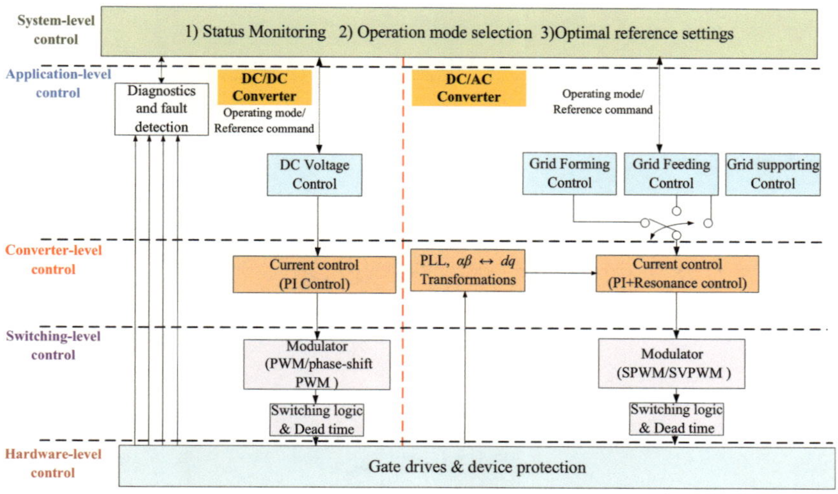

Figure 18.
Hierarchical control of power electronic converters.

design of the current controller is discussed. For the application-level control, control modes used for different applications are discussed.

4.1 Control of DC/DC converters

4.1.1 Switching-level control

PWM is the most used technique to control switching power devices in DC/DC converters. For example, to control a conventional buck DC/DC converter, a modulation wave $\boldsymbol{v}_m$ is generated from the control loop and compared with the sawtooth-wave carrier $\boldsymbol{v}_c$ as shown in **Figure 19**. The driving signal s (0 or 1) is sent to the driver according to (21).

$$\begin{aligned} &\text{If } v_m > v_c, s = 1 \\ &\text{If } v_m < v_c, s = 0 \end{aligned} \tag{21}$$

The larger the modulation wave, the larger the duty cycle, and thus the higher the output voltage. For other types of DC/DC converters such as the Dual-active-bridge (DAB) converter, the phase-shift PWM is favored to achieve zero-voltage-switching (ZVS) to reduce the losses [35].

4.1.2 Converter and application-level control

The DC/DC converters are used in the field of DG integration such as solar PV systems [36]. They transfer the power from DGs to DC microgrids. Types of DC/DC converters include buck, boost, and buck-boost converters.

Figure 20 shows a general control schematic for controlling the output voltage of a DC/DC converter. A double-loop controller is used for the DC/DC

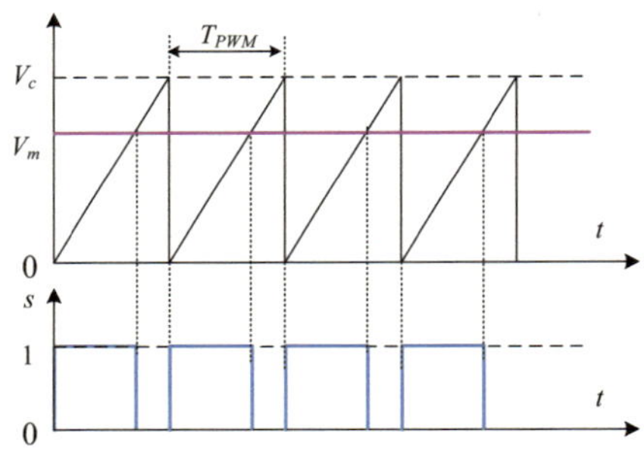

Figure 19.
PWM modulation scheme.

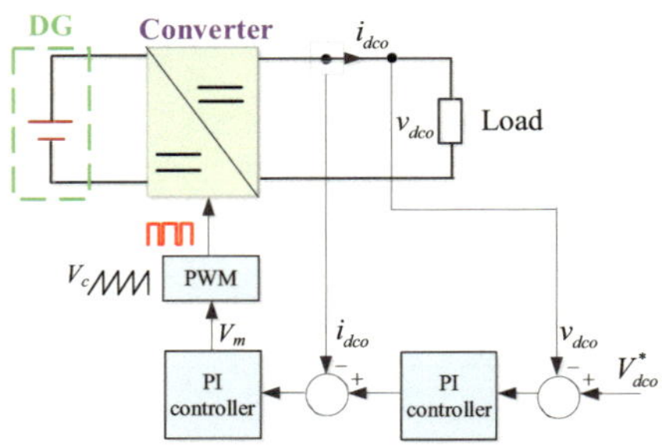

Figure 20.
Output DC voltage control.

converter. The control mode is to control the output DC voltage. The output from the voltage controller is the reference for the current controller. The modulation wave V_m is calculated from the current PI controller. Then, the PWM signal can be generated by the modulator as discussed in 4.1.1 to drive the power electronic switches.

For parallel-connected DC/DC converters in low-voltage DC microgrid. Droop control is also popular for DC/DC converters to achieve autonomous equal power-sharing. A virtual resistance R_V can be used to automatically distribute the power among the parallel converters. The droop curve and the control scheme are shown in **Figures 21** and **22**.

For PV energy integration, the DC/DC converter can be used for the MPPT control [37] as shown in **Figure 23**. A DC/DC converter is connected between the PV array and the load to trace the maximum powerpoint. In this case, the controller controls the input power through an MPPT algorithm. The output of the MPPT algorithm is a DC voltage reference. The input DC voltage of the DC/DC is then controlled according to this reference.

The load at the output side can be the passive load or the active load such as an AC/DC converter for connecting the PV system to an AC grid. In this case, the grid side AC/DC converter is responsible for regulating the DC link voltage.

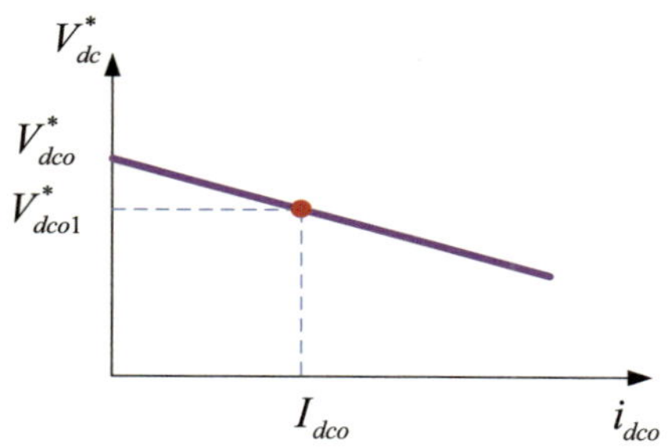

Figure 21.
Droop cure for virtual resistance-based control.

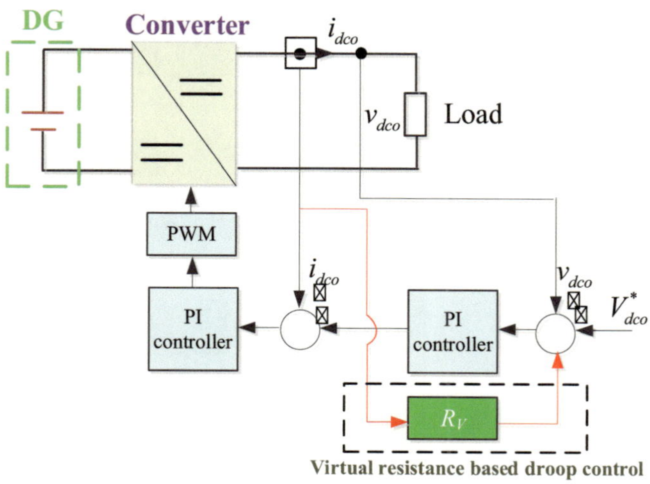

Figure 22.
Virtual resistance-based control schematic.

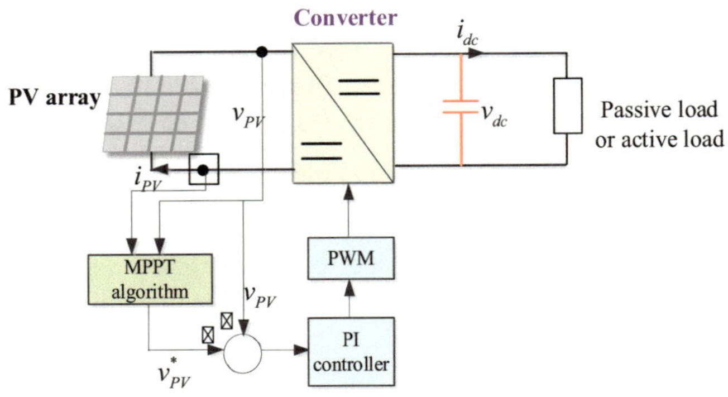

Figure 23.
Input DC voltage control for MPPT application.

4.2 Control of DC/AC converters

4.2.1 Switching-level control

Typical DC/AC converters used for connecting distributed generators (DGs) to the distribution networks are two-level (2 L) and three-level (3 L) converters, as shown in **Figure 24**.

Figure 25(a) and **(b)** show the sinusoidal PWM (SPWM) based modulation for 2 L and 3 L converters, respectively.

For the PWM modulation of 2 L converters in **Figure 25(a)**, the switching on and off of a IGBT device is determined by the comparison between the sinusoidal modulation wave (red dash line) and the carrier (gray triangle wave). For the modulation of 3 L converter in **Figure 25(b)**, the carrier disposition (CD) based modulation is commonly used. The CD modulation algorithm can be further

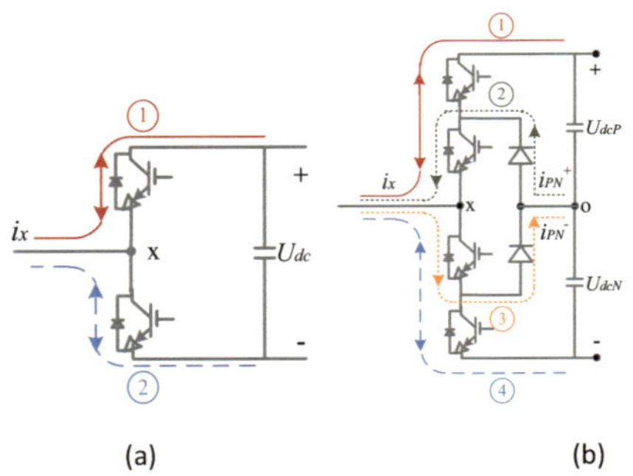

Figure 24.
SPWM for 2 L-converters, (a) a 2 L-converter, and (b) waveform of SPWM.

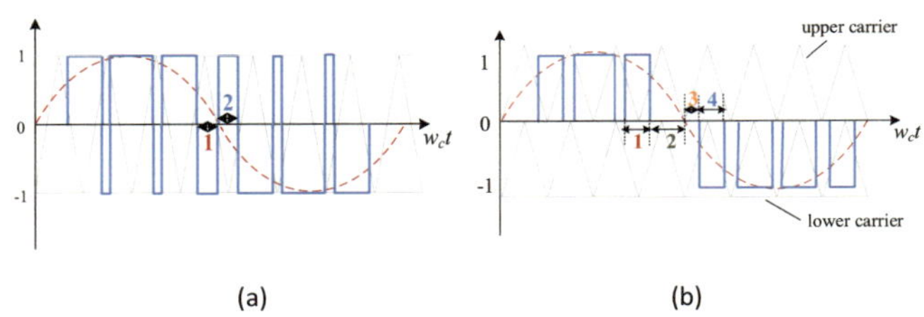

Figure 25.
SPWM for 3 L-converters, (a) a 3 L converter, and (b) waveform of SPWM.

categorized as phase disposition (PD) and phase opposite disposition (POD) methods. For the PD modulation, triangle carriers over and down the zero reference have the same phase (as shown in **Figure 25(b)**), whereas the phase of the lower carrier is opposite to the upper carrier for POD modulation. It can be observed that the phase voltage v_{xo} switches between $\frac{U_{dc}}{2}$ and 0 when the modulation wave (red dash line in **Figure 25(b)**) is positive, and between 0 and $-\frac{U_{dc}}{2}$ when modulation wave is negative.

4.2.2 Converter-level control

The decoupled current controller of the converter-level control is shown in **Figure 26**. Through the coordinate transformation from static frame (i.e. abc frame) to the synchronous rotating frame (i.e. dq frame), the active and reactive power is decoupled and can be controlled independently. Also, the control loop presents linear characteristics on the dq frame. Hence, the voltage-oriented vector control strategy is commonly adopted. The dq current controllers are based on PI regulators. The parameters of the PI regulators can be designed based on a second-order transfer function for a linear system. The decoupling terms $-\omega Li_q$ and ωLi_d are added to decouple the current components at the d-axis and q-axis. Voltage feedforward e_{dq} is used to improve the dynamic performance of the current controller.

For the grid-connected converters, the PLL is required to achieve synchronization to the grid frequency. The design of PLL in the $\alpha\beta$ frame is presented in **Figure 27**. The vector cross product is used to extract the error between the real phase angle of the grid voltage and the estimated phase voltage. The error passes through the PI regulator so that it can be eliminated.

The controller presented in **Figure 26** regulates positive-sequence current and works effectively if the voltage of the connected grid is balanced. However, if the voltage imbalance occurs, both positive-sequence and negative-sequence currents

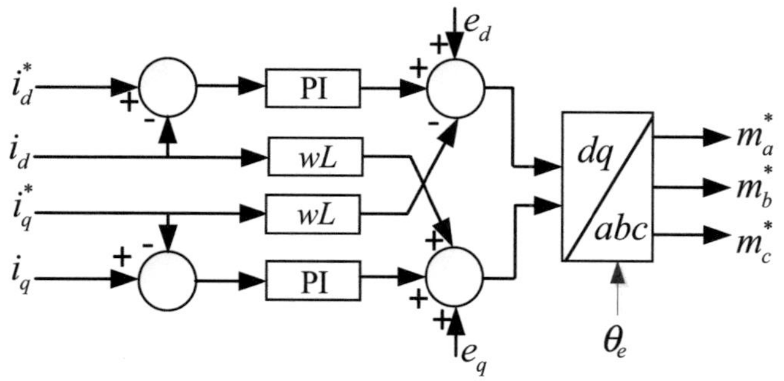

Figure 26.
Decoupled current controller at dq frames.

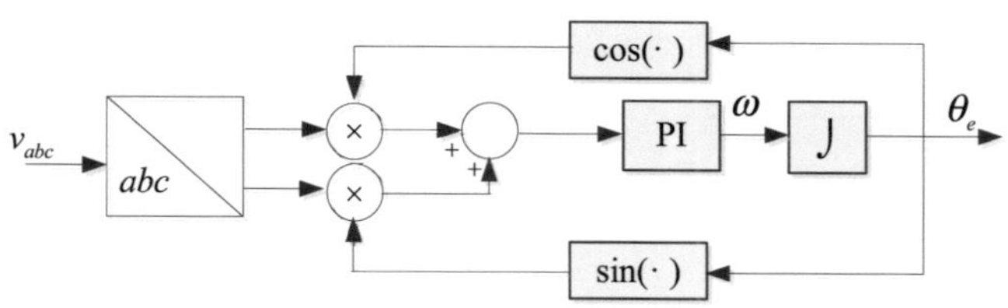

Figure 27.
PLL for grid synchronization.

will exist. Dual current controller regulating both positive-sequence and negative-sequence currents will be needed as shown in **Figure 28**.

The ultimate objective of using such dual current control to regulate currents of both sequences is to either i) achieve balanced output current [38], or ii) cancel the 2nd order power ripple caused by the interaction of positive-sequence current and negative-sequence voltage [39]. For i), balanced three-phase currents can be obtained by setting negative-sequence current references to zero. However, the interaction of positive-sequence current and negative-sequence voltage will result in a 2nd order power ripple at the grid side although the currents are controlled balanced through i):

$$\begin{aligned} p &= P_0 + P_{c2}\cos(2wt) + P_{s2}\sin(2wt) \\ q &= Q_0 + Q_{c2}\,cos\,(2wt) + Q_{s2}\,sin\,(2wt) \end{aligned} \tag{22}$$

This power ripple could increase odd AC harmonics to the grid. Thus, the approach for ii) is injecting proper negative-sequence current to counteract such a power ripple. The expected injecting currents are expressed as:

$$\begin{bmatrix} i_d^{p*} \\ i_q^{p*} \\ i_d^{n*} \\ i_q^{n*} \end{bmatrix} = \frac{2}{3}\begin{bmatrix} v_{sd}^p & v_{sq}^p & v_{sd}^n & v_{sq}^n \\ v_{sq}^p & -v_{sd}^p & -v_{sq}^n & v_{sd}^n \\ v_{sq}^n & -v_{sd}^n & -v_{sq}^p & v_{sd}^p \\ v_{sd}^n & v_{sq}^n & v_{sd}^p & v_{sq}^p \end{bmatrix}^{-1}\begin{bmatrix} P_0^* \\ Q_0^* \\ P_{s2}^* \\ P_{c2}^* \end{bmatrix} = \frac{2}{3}\begin{bmatrix} v_{sd}^p & v_{sq}^p & v_{sd}^n & v_{sq}^n \\ v_{sq}^p & -v_{sd}^p & -v_{sq}^n & v_{sd}^n \\ v_{sq}^n & -v_{sd}^n & -v_{sq}^p & v_{sd}^p \\ v_{sd}^n & v_{sq}^n & v_{sd}^p & v_{sq}^p \end{bmatrix}^{-1}\begin{bmatrix} P_0^* \\ Q_0^* \\ 0 \\ 0 \end{bmatrix} \tag{23}$$

Where, P_{s2}^* and P_{c2}^* are the sine and cosine terms of the 2nd order power ripples. In this scheme, the positive-sequence components $F_{\alpha\beta}^p = \begin{bmatrix} V_{\alpha\beta}^p & I_{\alpha\beta}^p \end{bmatrix}$ and the negative-sequence components $F_{\alpha\beta}^n = \begin{bmatrix} V_{\alpha\beta}^n & I_{\alpha\beta}^n \end{bmatrix}$ need to be extracted from $F_{\alpha\beta} = \begin{bmatrix} V_{\alpha\beta} & I_{\alpha\beta} \end{bmatrix}$. $F_{\alpha\beta}^p$ and $F_{\alpha\beta}^n$ can be obtained by misplaced subtraction as:

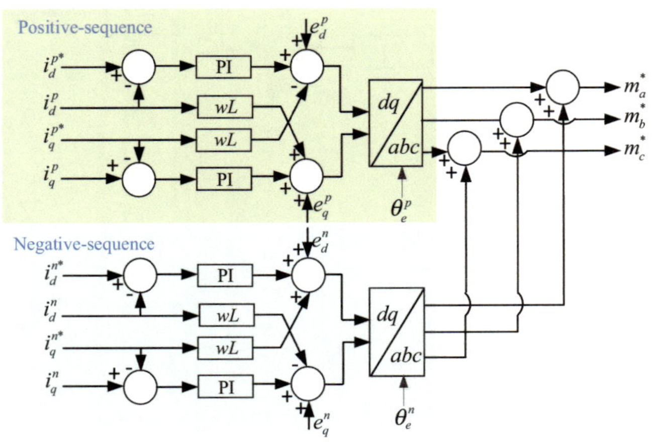

Figure 28.
Dual current controller for positive and negative sequence control.

$$\mathbf{F}^{\mathrm{p}}_{\alpha\beta}(\mathrm{t}) = \frac{1}{2}\left[\mathbf{F}_{\alpha\beta}(\mathrm{t}) + \mathbf{j}\mathbf{F}_{\alpha\beta}(\mathrm{t} - \mathrm{T}/4)\right]$$
$$\boldsymbol{F}^{n}_{\alpha\beta}(t) = \frac{1}{2}\left[\boldsymbol{F}_{\alpha\beta}(t) - j\boldsymbol{F}_{\alpha\beta}(t - T/4)\right] \tag{24}$$

Thus, the phase angles θ_e^{p} and θ_e^{n} in **Figure 28** can be obtained by phase locking the $\mathrm{V}^{\mathrm{P}}_{\alpha\beta}$ and $\mathrm{V}^{\mathrm{n}}_{\alpha\beta}$ separately.

In addition to the current controller in **Figure 26**, more control blocks can be added to reduce the harmonics that are generated by the nonlinear characteristics of converters. For example, carrier-based PWM methods can introduce the odd harmonics (i.e. 5th, 7th, 11th, and 13th), and the DC drift of neutral point voltage can cause even harmonics (i.e. 2nd and 4th). Therefore, the reduction of harmonics is required for converters. This can be achieved using the resonant (RES) controller [40]. The combined PI-RES controller at dq frames is shown in **Figure 29**.

A Nth order RES controller tuned at dq frames can compensate the (1-n)th and (1 + n)th harmonics in the stationary frame. For example, a 6th order RES controller can compensate the 5th and 7th harmonics. More RES controllers can be paralleled with the PI controller according to the compensating requirements.

4.2.3 Application-level control

At the application-level control, the control modes of the converters can be classified into three modes: grid-forming, grid-feeding, and grid-supporting modes [41]. The characteristics of the three modes are illustrated in **Figure 30**.

Converters in grid-forming mode should provide the AC voltage (reactive power) and frequency support (active power) for an AC grid. They act as voltage sources with low output impedance. The grid-forming converters can be also operated without being connected to the main grid.

Converters in grid feeding are operated as current sources and generate constant active and reactive power to the demand. However, as the operation of grid-feeding converters depends on at least one grid-forming converter, this type of converter cannot work without being connected to the main grid.

As for the microgrid, the power capacity of the DGs is limited. Thus, choosing one as the grid-forming converter may not be a good choice. Power-sharing

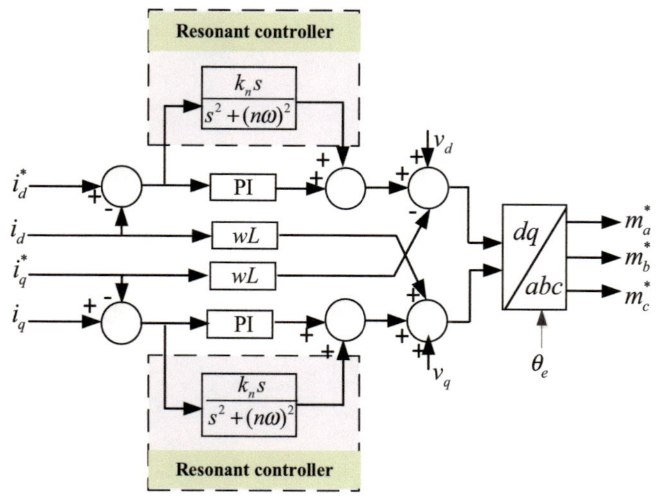

Figure 29.
PI-RES current controller.

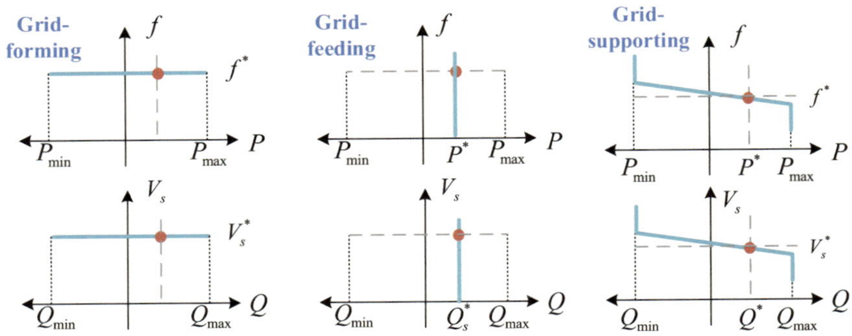

Figure 30.
Characteristics of three types of converter control.

methods to share the burden to each DG are preferred for the distribution networks integrated with renewable energies. To this end, the converters can work under grid-supporting mode and have a joint contribution to the voltage and frequency support. The basic power-sharing methods include centralized methods, master–slave methods, and distribution methods such as the droop methods [42]. The droop methods can automatically distribute the power to the DGs according to the droop curves, thus, communication is not required. Droop control for the grid-supporting converters is a promising power control strategy in the distribution network.

For the grid-forming converter control in **Figure 31**, the reference frequency f_0^* is given for the synchronization. The control is performed at dq frames. Feedback control is used to guarantee that the output voltage is equal to the given value. The outputs from the voltage PI controllers are the references for the current controller as mentioned in the previous section. The grid-forming mode is normally not used when being connected to the main grid, in which condition the grid-forming function is always performed by the synchronous generator of the power plants.

The control schematic for the grid-feeding converter is shown in **Figure 32**. The outer loop is the PI-based active and reactive power control loop. The active power and reactive power are calculated by the measured I_{abc} and V_{abc}. The outputs of the power control loops are the given values of the current control loops. Synchronization is required in this scheme, which is achieved using the PLL. The obtained phase angle θ_e is used for the transformation of voltage and current. The grid-feeding

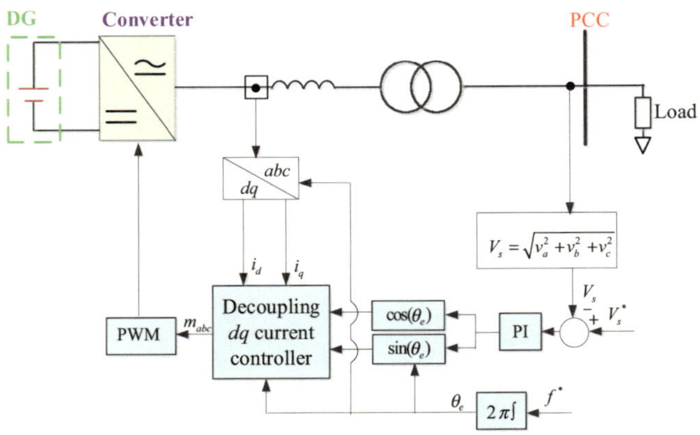

Figure 31.
Grid-forming control schematic.

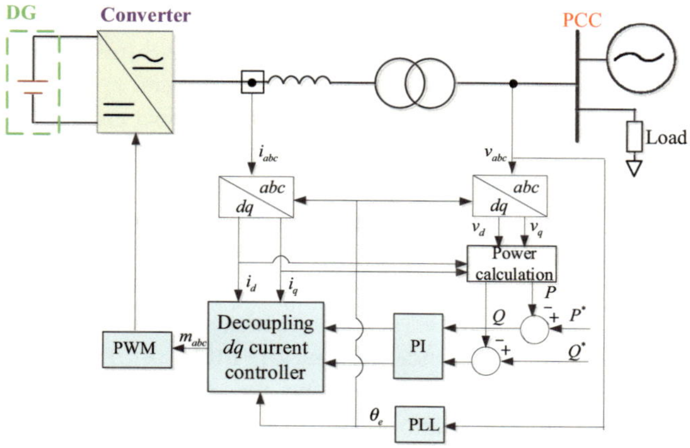

Figure 32.
Grid-feeding control schematic.

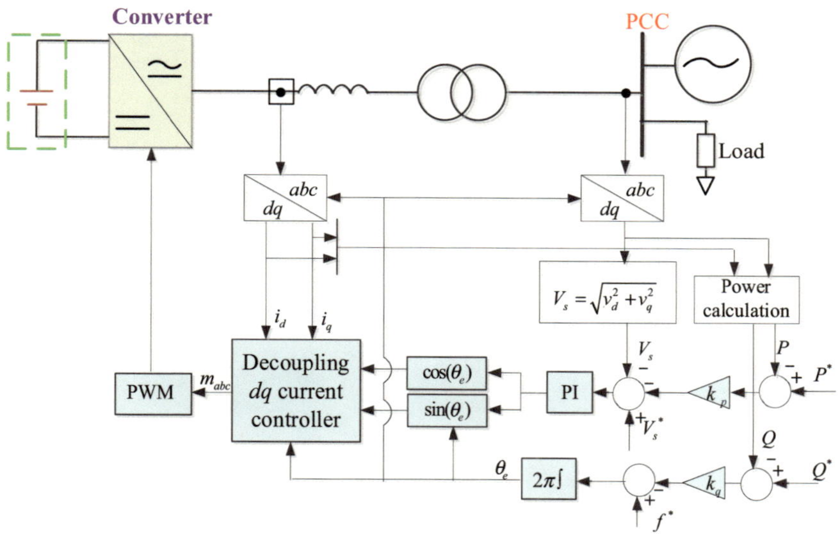

Figure 33.
Grid-supporting control schematic.

converter is always fed by the DGs which have more stochastic characteristics such as the PV and wind farms [41].

For the grid-supporting converter shown in **Figure 33**, the control scheme is the combination of the grid-forming and grid-feeding control methods. The $k_{p,q}$ is used to adjust the droop slope. The converter behaves more like a voltage source if $k_{p,q}$ increases. Otherwise, if $k_{p,q}$ decreases, the converter has more characteristics of the current source. The operation points (f^*, P^*) and (V^*, Q^*) in **Figure 33** are regulated by the secondary controller (system-level controller) [41].

5. Conclusion

Renewable energy resources, energy storage systems, and electric vehicles (EVs) are emerging in microgrids. A great many of the new energies are not naturally AC sources and cannot be connected to the grid. Power electronic converters build a bridge for the connection between renewable energies and microgrid.

Converter types include DC/DC and DC/AC converters. For the DC/DC converters, typical topologies are a buck, boost, and buck-boost converters. For the DC/AC converters, three-phase two-level and three-level converters are most widely used. In particular, the three-phase four-leg converters are used for the unbalanced load conditions. The neutral leg and current are independently controlled so that the dc capacitors can be choosed smaller to achieve lower cost and volume.

WBG devices have been used in the power converters. The design of WBG power converters should consider efficiency, volume, and weight, cost, and failure rate. To maximum these performance indices, the multi-optimization method is utilized. The optimized design solution is found according to the Pareto-front curve. To illustrate the design method, the optimization of efficiency and power density is particularly analyzed and validated through a prototype.

Control is an essential part of the power conversion. The typical control methods are discussed from the switching-level layer to the application-level layer following a hierarchical structure. PWM is a common modulation method. The current controllers are performed on the dq frame to achieve independent control of active and reactive power. Also, different control modes, such as grid-feeding, grid-forming, and grid-supporting control, are presented to accommodate different applications.

Author details

Wenlong Ming
The School of Engineering, Cardiff University, Cardiff, UK

*Address all correspondence to: mingw@cardiff.ac.uk

References

[1] Matos JG, Silva FSF, Ribeiro LAS. Power control in AC isolated microgrids with renewable energy sources and energy storage systems. IEEE Transactions on Industrial Electronics. 2015;**62**(6):3490-3498. DOI: 10.1109/TIE.2014.2367463

[2] Molina MG. Energy storage and power electronics technologies: A strong combination to empower the transformation to the smart grid. Proceedings of the IEEE. 2017;**105**(11): 2191-2219. DOI: 10.1109/JPROC.2017.2702627

[3] Rahman MS, Hossain MJ, Lu J, Pota HR. A need-based distributed coordination strategy for EV storages in a commercial hybrid AC/DC microgrid with an improved interlinking converter control topology. IEEE Transactions on Energy Conversion. 2018;**33**(3):1372-1383. DOI: 10.1109/TEC.2017.2784831

[4] Rocabert J, Luna A, Blaabjerg F, Rodríguez P. Control of power converters in AC microgrids. IEEE Transactions on Power Electronics. 2012;**27**(11):4734-4749. DOI: 10.1109/TPEL.2012.2199334

[5] Dallago E, Liberale A, Miotti D, Venchi G. Direct MPPT algorithm for PV sources with only voltage measurements. IEEE Transactions on Power Electronics. 2015; **30**(12):6742-6750. DOI: 10.1109/TPEL.2015.2389194

[6] Abo-Khalil AG, Lee D. MPPT control of wind generation systems based on estimated wind speed using SVR. IEEE Transactions on Industrial Electronics. 2008;**55**(3):1489-1490. DOI: 10.1109/TIE.2007.907672

[7] Long C, Wu J, Zhou Y, Jenkins N. Peer-to-peer energy sharing through a two-stage aggregated battery control in a community Microgrid. Applied Energy. 2018;**226**:261-276. DOI: 10.1016/j.apenergy.2018.05.097

[8] Zhang C, Wu J, Zhou Y, Cheng M, Long C. Peer-to-peer energy trading in a microgrid. Applied Energy. 2018;**220**:1-12. DOI: 10.1016/j.apenergy.2018.03.010

[9] Li J, Liu Y, Wu L. Optimal operation for community-based multi-party microgrid in grid-connected and islanded modes. IEEE Transactions on Smart Grid. 2018;**9**(2):756-765. DOI: 10.1109/TSG.2016.2564645

[10] Shen X, Tan D, Shuai Z, Luo A. Control techniques for bidirectional interlinking converters in hybrid microgrids: Leveraging the advantages of both AC and DC. IEEE Power Electronics Magazine. 2019;**6**(3):39-47. DOI: 10.1109/MPEL.2019.2925298

[11] Che L, Shahidehpour M, Alabdulwahab A, Al-Turki Y. Hierarchical coordination of a community microgrid with AC and DC microgrids. IEEE Transactions on Smart Grid. 2015;**6**(6):3042-3051. DOI: 10.1109/TSG.2015.2398853

[12] Mandrile F, Carpaneto E, Bojoi R. Grid-feeding inverter with simplified virtual synchronous compensator providing grid services and grid support. IEEE Transactions on Industry Applications. 2021;**57**(1):559-569. DOI: 10.1109/TIA.2020.3028334

[13] Yazdani S, Ferdowsi M, Davari M, Shamsi P. Advanced current-limiting and power-sharing control in a PV-based grid-forming inverter under unbalanced grid conditions. IEEE Journal of Emerging and Selected Topics in Power Electronics. 2020;**8**(2):1084-1096. DOI: 10.1109/JESTPE.2019.2959006

[14] Amjadi Z, Williamson SS. Power-electronics-based solutions for plug-in hybrid electric vehicle energy storage and management systems. IEEE Transactions on Industrial Electronics. 2010;**57**(2):608-616. DOI: 10.1109/TIE.2009.2032195

[15] Hmad J, Houari A, Trabelsi H, Machmoum M. Fuzzy logic approach for smooth transition between grid-connected and stand-alone modes of three-phase DG-inverter. Electric Power Systems Research. 2019;**175**:105892. DOI: 10.1016/j.epsr.2019.105892

[16] Jiang Q, Xue M, Geng G. Energy management of microgrid in grid-connected and stand-alone modes. IEEE Transactions on Power Systems. 2013; **28**(3):3380-3389. DOI: 10.1109/TPWRS.2013.2244104

[17] Mirjafari M, Harb S, Balog RS. Multiobjective optimization and topology selection for a module-integrated inverter. IEEE Transactions on Power Electronics. 2015;**30**(8): 4219-4231. DOI: 10.1109/TPEL.2014.2353055

[18] Burkart RM, Kolar JW. Comparative life cycle cost analysis of Si and SiC PV converter systems based on advanced η-ρ-σ multiobjective optimization techniques. IEEE Transactions on Power Electronics. 2017;**32**(6):4344-4358. DOI: 10.1109/TPEL.2016.2599818

[19] Villar-Piqué G, Bergveld H, Alarcón E. Survey and benchmark of fully integrated switching power converters: Switched-capacitor versus inductive approach. IEEE Transactions on Power Electronics. 2013;**28**(9):4156-4167. DOI: 10.1109/TPEL.2013.2242094

[20] Tasi-Fu W, Yu-Kai C. Modeling PWM DC/DC converters out of basic converter units. IEEE Transactions on Power Electronics. 1998;**13**(5):870-881. DOI: 10.1109/63.712294

[21] Mohan N, Underland TM, Robbins WP. Power Electronics Converters, Applications and Design. 3rd ed. Wiley; 2003. 802 p. ISBN: 978-0-471-22693-2

[22] Hart DW. Power Electronics. 1st ed. Mc Graw Hill; 2011. 477 p. ISBN 978-0-07-338067-4

[23] Kim K, Cha H, Kim HG. A new single-phase switched-coupled-inductor DC–AC inverter for photovoltaic systems. IEEE Transactions on Power Electronics. 2017;**32**(7):5016-5022. DOI: 10.1109/TPEL.2016.2606489

[24] Liu B, Wang L, Song D, Su M, Yang J, He D, et al. Input current ripple and grid current harmonics restraint approach for single-phase inverter under battery input condition in residential photovoltaic/battery systems. IEEE Transactions on Sustainable Energy. 2018;**9**(4): 1957-1968. DOI: 10.1109/TSTE.2018.2820507

[25] Lee DC, Kim YS. Control of single-phase-to-three-phase AC/DC/AC PWM converters for induction motor drives. IEEE Transactions on Industrial Electronics. 2007;**54**(2):797-804. DOI: 10.1109/TIE.2007.891780

[26] Jeong H-G, Kim G-S, Lee K-B. Second-order harmonic reduction technique for photovoltaic power conditioning systems using a proportional-resonant controller. MDPI Solar Energy Systems and Materials. 2013;**6**(1):79-96. DOI: 10.3390/en6010079

[27] Boglietti A, Bojoi R, Cavagnino A, Tenconi A. Efficiency analysis of PWM inverter fed three-phase and dual three-phase high frequency induction machines for low/medium power applications. IEEE Transactions on Industrial Electronics. 2008;**55**(5): 2015-2023. DOI: 10.1109/TIE.2008.918489

[28] Ahmed MH, Wang M, Hassan MAS, Ullah I. Power loss model and efficiency analysis of three-phase inverter based on SiC MOSFETs for PV applications. IEEE Access. 2019;7:75768-75781. DOI: 10.1109/ACCESS.2019.2922741

[29] Dai NY, Wong MC, Han YD. Application of a three-level NPC

inverter as a three-phase four-wire power quality compensator by generalized 3DSVM. 2006;**21**(2): 440-449. DOI: 10.1109/ TPEL.2005.869755

[30] Sun M, Demirtas S, Sahinoglu Z. Joint voltage and phase unbalance detector for three phase power systems. IEEE Signal Processing Letters. 2013; **20**(1):11-14. DOI: 10.1109/ LSP.2012.2226717

[31] Liang J, Green TC, Feng C, Weiss G. Increasing voltage utilization in split-link, four-wire inverters. IEEE Transactions on Power Electronics. 2009;**24**(6):1562-1569. DOI: 10.1109/ TPEL.2009.2013351

[32] Fu Y, Huang Y, Lu X, Zou K, Chen C, Bai H. Imbalanced regulation based on virtual resistance of a three-phase four-wire inverter for EV vehicle-to-home applications. IEEE Transactions on Transportation Electrification. 2019; **5**(1):162-173. DOI: 10.1109/ TTE.2018.2874357

[33] Alex L, Michael DR, Johan S, David R, John G. GaN Transistors for Efficient Power Conversion. John Wiley & Sons; 2019

[34] Stupar A, Friedli T, Minibock J, Kolar JW. Towards a 99% efficient three-phase buck-type pfc rectifier for 400-v dc distribution systems. IEEE Transactions on Power Electronics. 2012;**27**(4):1732-1744

[35] Taylor A, Liu G, Bai H, Brown A, Johnson PM, McAmmond M. Multiple-phase-shift control for a dual active bridge to secure zero-voltage switching and enhance light-load performance. IEEE Transactions on Power Electronics. 2018;**33**(6):4584-4588. DOI: 10.1109/TPEL.2017.2769638

[36] Li W, He X. Review of nonisolated high-step-up DC/DC converters in photovoltaic grid-connected applications. IEEE Transactions on Industrial Electronics. 2011;**58**(4): 1239-1250. DOI: 10.1109/ TIE.2010.2049715

[37] de Brito MAG, Galotto L, Sampaio LP, Melo GDAE, Canesin CA. Evaluation of the main MPPT techniques for photovoltaic applications. IEEE Transactions on Industrial Electronics. 2013;**60**(3): 1156-1167. DOI: 10.1109/ TIE.2012.2198036

[38] Joseph T, Liang J, Li G, Moon A, Smith K, Yu J. Dynamic control of MVDC link embedded in distribution network: — Case study on ANGLE-DC. In: 2017 IEEE Conference on Energy Internet and Energy System Integration (EI2). 2017. pp. 1-6. DOI: 10.1109/ EI2.2017.8245752

[39] Tao W, Zhixia G, Wang L, Li J. Research on control strategy of grid-connected inverter under unbalanced voltage conditions. In: 2016 IEEE 8th International Power Electronics and Motion Control Conference (IPEMC-ECCE Asia). 2016. pp. 915-919. DOI: 10.1109/IPEMC.2016.7512408

[40] Liserre M, Teodorescu R, Blaabjerg F. Multiple harmonics control for three-phase grid converter systems with the use of PI-RES current controller in a rotating frame. IEEE Transactions on Power Electronics. 2006;**21**(3):836-841. DOI: 10.1109/TPEL.2006.875566

[41] Rocabert J, Luna A, Blaabjerg F, Rodríguez P. Control of power converters in AC microgrids. IEEE Transactions on Power Electronics. 2012;**27**(11):4734-4749. DOI: 10.1109/ TPEL.2012.2199334

[42] Su W, Huang AQ. The Energy Internet: An Open Energy Platform to Transform Legacy Power Systems Into Open Innovation and Global Economic Engines. Duxford, UK: Woodhead Publishing; 2019. pp. 123-152

Chapter 3

Peer-to-Peer Energy Trading in Microgrids and Local Energy Systems

Yue Zhou and Jianzhong Wu

Abstract

Peer-to-peer (P2P) energy trading is an innovative approach for managing increasing numbers of Distributed Energy Resources in microgrids or local energy systems. In P2P energy trading, prosumers and consumers directly trade and exchange power and energy with each other. The development of P2P energy trading is described in five key aspects, that is, market design, trading platforms, power and ICT infrastructure, regulation and policy, and from a social science perspective. A general multiagent framework is established to simulate the behaviour of and interaction between multiple entities in P2P energy trading. A general evaluation index hierarchy is proposed to assess various P2P energy trading mechanisms. Finally, a residential community that is set in the context of Great Britain is studied using multiagent simulation and hierarchical evaluation methods. Both the technical and economic benefits of P2P energy trading are demonstrated.

Keywords: Peer-to-peer energy trading, Distributed energy resource, Local energy system, Microgrid

1. Introduction

Traditional electric power systems operate in a unidirectional way, i.e. electricity is generated by large centralised generators, transmitted through transmission and distribution networks, and finally delivered to end users. Accordingly, electricity is sold through a wholesale market and large centralised generators sell electricity in bulk to energy suppliers (sometimes called 'retailers'), who will further re-sell it in small amounts to end users.

However, the connection of Distributed Energy Resources (DERs) changes the landscape radically. DERs include distributed generators (especially renewable energy generators such as solar panels and wind generators), energy storage systems and flexible demand. DERs may cause reversed power flow, and both power systems and electricity markets become bidirectional. End users, who have the capability of generating electricity on site (called 'prosumers', which is a word combining 'producer' and 'consumer'), are able to feed electricity back to the bulk power grid and obtain payment from energy suppliers.

Typical schemes to remunerate power fed from renewable sources into the bulk power grid are 'net metering' and 'Feed-in Tariff (FiT)' schemes. In net metering, the surplus generation that is fed back to the bulk power grid is recorded and then deducted from the electricity consumption of the same prosumer over a period of time (e.g., a month or a year). In the FiT scheme, the generation that is fed into the bulk power grid is remunerated at a fixed export price.

However, with rapidly increasing connection of DERs, net metering or FiT schemes impose financial pressure on the power utilities and any customers who do not install DERs. As a consequence, many countries are reducing their support for distributed renewable generation through these schemes. In this context, prosumers are seeking alternatives to selling surplus generation to the bulk power grid. Peer-to-peer (P2P) energy trading is an emerging solution, where prosumers and consumers exchange energy with each other in microgrids or Local Energy Systems (LES) [1].

2. Key aspects of peer-to-peer energy trading

As a promising scheme to tackle high penetration of DERs, P2P energy trading has attracted increasing attention and investment in many countries of the world. A rapidly increasing number of academic papers, research projects and industrial practice have emerged on P2P energy trading around the world [2]. P2P energy trading is a comprehensive scheme involving multiple spatial scales, multiple time scales and multiple conceptual layers, as summarised and illustrated in **Figure 1**.

Conceptual layers

P2P energy trading is a broad concept, beyond the power engineering domain. To implement P2P energy trading, various components are needed, including the electricity infrastructure to physically deliver the electricity traded, ICT infrastructure for supporting information exchange and advanced control, trading platforms for participants to negotiate and trade with each other, market design to set the trading rules, and laws and policies to regulate and guide the trading. Social science perspective is also valuable, which can help better understand participants' behaviours and situations to improve the design and fairness in P2P energy trading.

Spatial scales

As shown in **Figure 1**, spatially, a hierarchical structure can be established for P2P energy trading in power systems. At the bottom level, prosumers and consumers can trade with each other within a microgrid/LES (e.g., in a community area). However, the generation and demand might not completely match with each other, resulting in the potential needs of the trading between microgrids/LES to consume/supply each other's energy surplus/deficit. If the power and energy

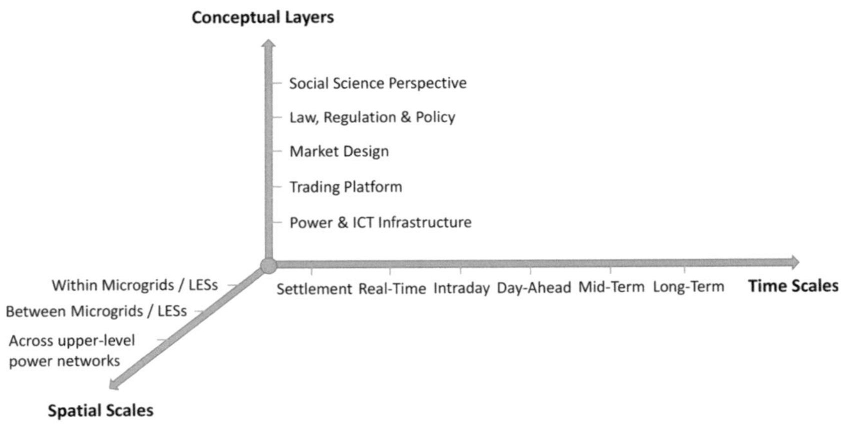

Figure 1.
P2P energy trading with multiple spatial scales, time scales and conceptual layers.

still cannot be balanced, electricity needs to be imported from / exported to upper-level power networks. With this hierarchical P2P energy trading structure, power and energy can be balanced layer by layer in a bottom-up manner, so that the capacity and losses of electricity transmission equipment can be reduced. In this chapter, we focus on P2P energy trading between prosumers and consumers in microgrids and LES.

Time scales

P2P energy trading spans a wide range of time scales as well. Similar to traditional energy trading in the electricity wholesale market, trading contracts can be made from well in advance (i.e., long-term/mid-term trading, such as year/month-ahead), day-ahead, intraday, to real time. After the delivery time, settlement needs to be conducted to examine whether and to what extent the participants in P2P energy trading have followed the pre-made contracts, and then to execute the financial payment/penalty accordingly.

2.1 Market design

Market design specifies the rules that the participants must follow to conduct energy trading with each other, including the requirements for information to be provided by the participants (e.g., the bids of the amount of energy and price in an auction), the rules to match generation and demand, the pricing model and the market settlement mechanism.

Currently, market design has attracted attention from both academia and industry [2]. Many market concepts have been proposed, which can be largely classified into three categories, i.e., centralised, decentralised and distributed markets. In a *centralised market*, there will be one central market coordinator, which collects all the information needed from prosumers and consumers, conducts the matching and pricing centrally, and settles the market or even directly controls some devices of prosumers/consumers. Centralised markets are relatively easy to design, being able to maximise the social welfare of all the participants and having higher level of certainty in terms of the participants' behaviours. However, centralised markets are vulnerable to single-point failures, cause potential concerns over participants' privacy and autonomy, and have scalability issues if the number of participants in the markets increases significantly.

The opposite type of markets is the *decentralised* ones, in which there is not a central market coordinator, and all the prosumers and consumers directly negotiate and trade with each other bilaterally, e.g., establishing a 'bilateral contract network' [3]. In contrast to centralised markets, decentralised markets protect participants' privacy and autonomy and have good scalability, but it is more difficult to reach a stable outcome which maximises the social welfare.

The compromise solution is *distributed* markets, where there is still a central market coordinator, but collecting less information from the participants and intervening with the participants in a more indirect way (e.g., through pricing signals rather than direct control signals). A number of distributed market designs are based on the Stackelberg game [4] or Alternating Direction Method of Multipliers (ADMM) [5]. Generally, distributed markets combine the advantages and disadvantages of centralised and decentralised markets.

Although many mechanisms have been proposed, there are still many challenges in the market design of P2P energy trading. Prosumers and consumers can be seen as profit-driven entities, aiming to maximum profits / minimum costs in P2P energy trading, but they often have conflicting interests, and thus how to model and manage the complicated interaction among prosumers and consumers is a challenge. Game-theory methods, either cooperative or non-cooperative games, can be used to deal with this challenge [6]. Also, if multiple P2P energy trading markets are allowed in the same area, the dynamics of forming and dissolving P2P energy trading coalitions becomes an interesting and practical topic, which has not been well explored yet. Furthermore, with increasing adoption of P2P energy trading in the future, the relationship between P2P energy trading markets and existing electricity wholesale and retail markets needs to be re-defined and coordinated. The current assumption that P2P energy trading markets are just price takers of bulk retail / wholesale markets will no longer be the case (**Box 1**).

The Mid-Market Rate (MMR) mechanism, first proposed in [7], is used as an example to demonstrate P2P energy trading market design. MMR is a centralised market, where a central market coordinator is needed for information collection, pricing and settlement. Spatially, the MMR mechanism operates within a community microgrid. Temporally, MMR mechanism applies to time scales from day-ahead to hour-ahead energy trading.

Consider an example microgrid with only one prosumer *A* and one consumer *B*.

First, assume *A* has surplus generation of P_A = 5kWh to be sold, and *B* has the same amount of demand, i.e. P_B = 5kWh, to be supplied in a certain time slot.

Conventionally, *A* and *B* are individually metered and trade with the energy supplier. Assume the energy price at which prosumers/consumers buy electricity is p_{buy} = 15 p/kWh and the price to sell electricity to the energy supplier is p_{sell} = 5 p/kWh,. Then the net costs of them are (positive for cost, negative for income)

$$C_{A\_CONV} = -p_{sell} \times P_A = -5 \times 5 = -25\,p \tag{1}$$

$$C_{B\_CONV} = p_{buy} \times P_B = 15 \times 5 = 75\,p \tag{2}$$

By contrast, in the MMR mechanism, *A* and *B* submit their surplus generation and electricity demand to the market coordinator, which makes the price for P2P energy trading as

$$p_{P2P} = \frac{p_{buy} + p_{sell}}{2} = \frac{15+5}{2} = 10\,p/kWh. \tag{3}$$

The demand of *B* will be completely supplied by the surplus generation of *A* at p_{P2P}. Therefore, the net costs of *A* and *B* in the MMR mechanism will be

$$C_{A\_P2P} = -p_{P2P} \times P_A = -10 \times 5 = -50\,p, \tag{4}$$

$$C_{B\_P2P} = p_{P2P} \times P_B = 10 \times 5 = 50\,p. \tag{5}$$

The foundation of the MMR mechanism is the assumption that p_{buy} is larger than p_{sell}, which is usually the case in most countries (e.g., in Great Britain, p_{buy} is about three times as high as p_{sell}). With this assumption, we can see that $|C_{A\_\text{P2P}}| > |C_{A\_\text{CONV}}|$ while $|C_{B\_\text{CONV}}| > |C_{B\_\text{P2P}}|$, indicating that the prosumer *A* has higher income while the consumer *B* has lower cost in P2P energy trading, so both benefit. The MMR mechanism takes the P2P energy trading price as the mean of the buying and selling prices issued by the energy supplier.

Box 1.
An example of P2P energy trading market design – Mid-market rate (MMR) mechanism.

2.2 Trading platforms

Once proper design of P2P energy trading markets is in place, trading platforms are needed for the prosumers, consumers and coordinators to exchange information and negotiate with each other, make deals and transactions, and conduct other relevant activities such as problem reporting and dispute resolution. Considering that trading frequencies are usually high in P2P energy trading, the trading platforms are usually web-based services, which can be accessed by the participants through smart phones, tablets or personal computers in a convenient way.

A trading platform can be built based either on conventional centralised servers and databases or on Distributed Ledger Technology (DLT). Blockchain, which is the underlying technology of Bitcoin and many other cryptocurrencies, is the most famous and widely used type of DLT. Blockchain has a number of features including trustworthiness, transparency, redundancy, tamper-proof ability, and intermediary avoidance, which means it has good potential to be used for supporting P2P energy trading. Furthermore, smart contracts, which are contracts in the form of computer codes that can be automatically executed, can reduce the costs of contracting, enforcement and compliance in P2P energy trading [8].

There have been many industrial projects using Blockchain for P2P energy trading, such as the Brooklyn microgrid case in the U.S. [9]. In spite of the many potential advantages and its increasing use, there are also many issues of using Blockchain or wider DLT for P2P energy trading. DLT is still young and undergoing rapid development, with a number of technical issues, e.g., the high computational and energy costs of the 'Proof of Work (PoW)' mechanism and the reduced level of trustworthiness of other consensus mechanisms (**Box 2**).

A software platform for P2P energy trading, named as "Elecbay", has been proposed in [1] and introduced here as an example. The parties and their interaction within "Elecbay" are illustrated in **Figure 2**.

The design of "Elecbay" learns from that of eBay, which is a famous e-commercial platform enabling customer-to-customer or business-to-customer sales online. On "Elecbay", electricity sellers list the electricity to be sold as items, while electricity buyers browse all the listed items and place orders. Each order specifies the amount, time duration and price of the electricity to be exchanged.

After the transactions are agreed between electricity buyers and sellers, they will be sent to network operators to check whether the orders satisfy physical network constraints. Network operators can reject the agreements that will cause problems to physical networks. Detailed mechanisms in this will be discussed in Section 2.3.

The electricity to be sold by the sellers may not be enough to satisfy all the needs of the buyers, and vice versa. Therefore, electricity suppliers cover this electricity imbalance (i.e., providing balancing service) through buying/selling electricity from/to sellers/buyers.

Finally, at the settlement stage, "Elecbay" will distribute the payment collected from electricity buyers among electricity sellers (for providing the electricity), network operators (for the use of networks) and electricity suppliers (for dealing with the electricity imbalance). The "Elecbay" platform may keep a small percentage of the payment as the service fee as well.

Box 2.
"Elecbay" – An example of P2P energy trading platform.

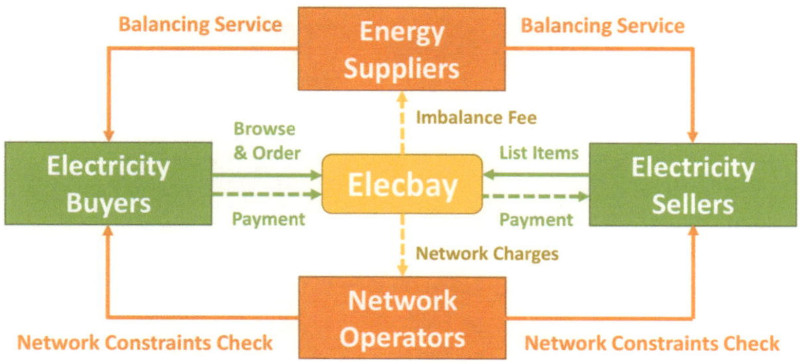

Figure 2.
The parties and their interaction within "Elecbay".

2.3 Power and ICT infrastructure

The transactions agreed between prosumers and consumers in P2P energy trading need to be physically delivered through electric power networks. This can be private wires or public power networks, as illustrated in **Figure 3**.

If a prosumer and a consumer (or two prosumers) are connected purely by private wires, as shown in **Figure 3(a)**, the situation is simple and clear, where the electricity can be transmitted just as agreed in P2P energy trading. However, this is a rare scenario in the real world – electricity transmission and distribution businesses benefit from economies of scale and economies of density, so it is generally not economic (and even illegal in some areas) to build private power networks. Typical examples of private power networks include those in industrial parks, newly built residential communities and microgrids in islands or remote areas.

Therefore, in most practical cases, prosumers and consumers are interconnected with each other through public power networks, as shown in **Figure 3(b)**. Electricity in public power networks has unified power quality, and the electricity from different sources cannot be physically distinguished. In this case, P2P energy trading is in nature a virtual trading arrangement, similar to the bilateral electricity purchase contracts between larger generators and consumers in conventional electricity markets. In spite of the similarities, a lot of new commercial mechanisms need to be set up to allocate the network use charges and network losses, manage the operational constraints, and provide incentives for long term network investment.

Information and communication technology (ICT) infrastructure is also needed to achieve P2P energy trading. Metering infrastructure is needed for measuring the generation/consumption of prosumers and consumers in P2P energy trading for market settlement and billing purposes. The smart metering infrastructure being deployed in many areas of the world is useful in this regard. There is research showing that existing communication technologies are sufficient to support P2P energy trading [10, 11].

2.4 Law, regulation, and policy

There have not been many formal and widely applied policies, laws or regulations on P2P energy trading across the world. P2P energy trading and the corresponding supporting technologies like Blockchain are still at an early stage and under rapid development, so it may still be too early to make firm and widely applied policies. Trials that rely on derogations from existing regulations have the advantages of being

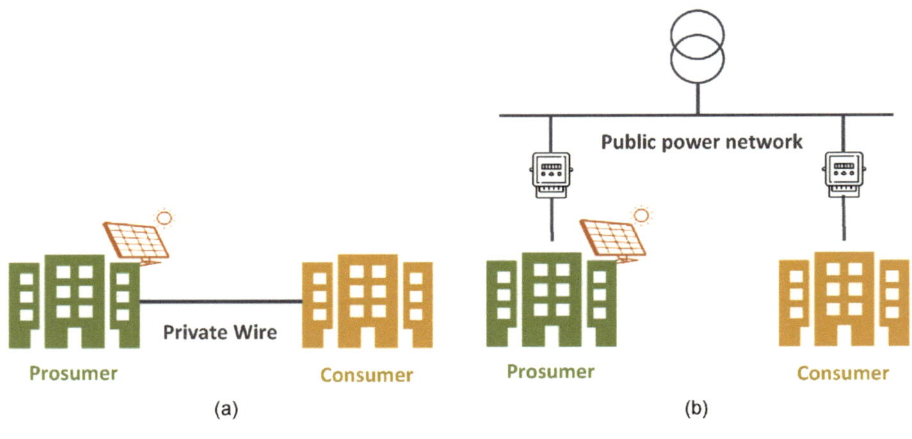

Figure 3.
P2P energy trading through private wires and public power networks. (a) through private wires. (b) through public power networks.

very flexible and relatively easy to establish. However, with the rapidly increasing penetration of DERs and increasing scale of P2P energy trading, establishing systematic policy, regulation and legal frameworks will be required (**Box 3**).

The following issues need to be addressed in future policy development:

- the formal role and responsibilities of participants of P2P energy trading (e.g., prosumers, consumers and the P2P energy trading coordinators),
- the relationship between P2P energy trading and existing electricity markets (e.g., wholesale market, retail market, capacity market and ancillary service market),
- the distribution of levies, taxes and network charges among the participants of P2P energy trading,
- appropriate subsidies and incentives for encouraging the development of P2P energy trading, and
- protection of vulnerable customers and energy equity in P2P energy trading.

Box 3.
A list of issues to be addressed in future policy development.

2.5 A social science perspective

P2P energy trading involves large numbers of small customers, who cannot be treated in the same way as large generation companies or electricity suppliers, which behave with almost perfectly economical rationality and have a high-risk tolerance. Therefore, a social science perspective is of great importance in P2P energy trading to design reasonable and effective market mechanisms and to better understand, engage, satisfy, and protect customers (**Box 4**).

1. Is P2P energy trading all settled by cash?
   No! In an ethnographic study in two off-grid villages in rural India, Singh et al. found that P2P energy trading is not just an economic act but actually a complicated sociocultural process [12]. Besides the 'in-cash' return (i.e., money), there are 'in-kind' return (e.g., food, oil, cakes, and service of irrigation) and 'intangible' returns (e.g., goodwill and friendship), involving the dynamics of social relations.

2. Does P2P energy trading always happen in a marketplace?
   No! In another ethnographic study conducted in an off-grid village in rural India, Singh et al. found that P2P energy trading can not only happen in the regulated market realm mainly with rational participants, but also as 'a social and personal transaction of energy between giver and energy receiver'. This was named as 'mutual energy trading', which is mutually structured and negotiated [13].

3. Do customers value 'autarky' or 'autonomy' more?
   Based on the result of an online survey involving 248 German homeowners, Ecker et al. found that most customers value 'autarky' (i.e., independence of power supply) more than 'autonomy' (i.e., the ability to self-determine the source of energy) [14]. Ecker et al. inferred that this preference might not be good news for deploying P2P energy trading, considering customers might price the electricity they generate at a higher price than its actual worth. This inference remains to be validated in practice.

Box 4.
Example social science questions in P2P energy trading.

3. Simulation and evaluation of peer-to-peer energy trading

More and more P2P energy trading mechanisms with various focuses and designs are being created by researchers and practitioners across the world. Therefore, instead of focusing on any specific P2P energy trading mechanism which might be upgraded or replaced quickly, a general simulation and evaluation framework of P2P energy trading is described. This framework facilitates an understanding of the interaction of various parties, and can be used to simulate and assess the outcome of P2P energy trading for improving the mechanism design and conducting feasibility analysis and impact assessment.

3.1 Multiagent simulation

P2P energy trading involves multiple parties with different interests and objectives, and thus is suitable to be modelled and simulated by a multiagent system. The overall picture of the multiagent simulation framework for P2P energy trading is shown in **Figure 4**.

Figure 4 shows that the framework models the parties related to P2P energy trading as agents and also describes their interactions. In the lower block, the parties within the P2P trading community and their interactions are presented. Although the term 'community' is used, it is not necessary to be a physical community, but can be a virtual aggregation of the parties conducting P2P energy trading. The participants of P2P energy trading include prosumers, distributed generators (e.g., small-scale PV power plants), community energy storage (e.g., batteries collectively owned and shared by the customers in a community) and consumers. There is a coordinator managing various aspects of P2P energy trading. This trading coordinator could be a real entity, e.g., a company running this business, or a virtual one, such as a smart contract sitting in a blockchain. Depending on the specific P2P energy trading mechanisms, the participants need to exchange various types of information (such as measurements, bids and offers, pricing signals, and control

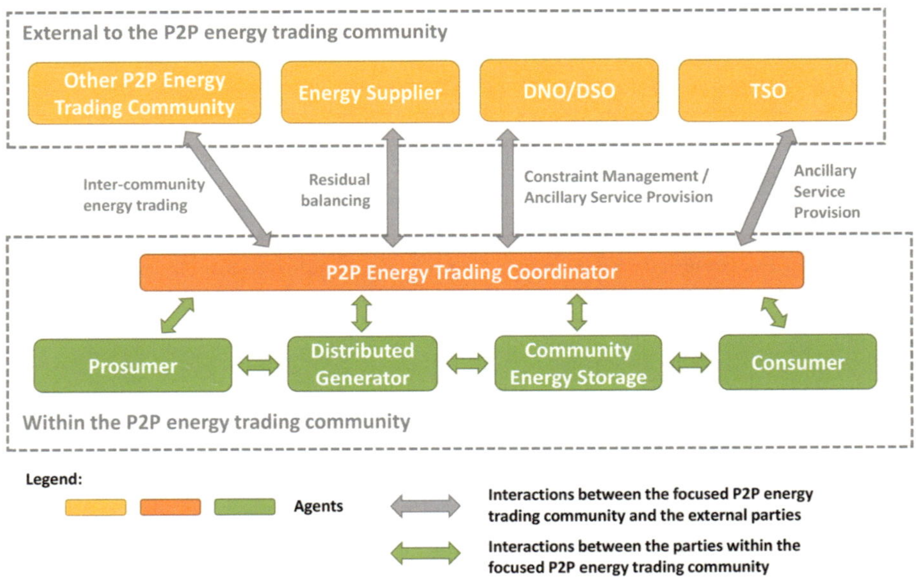

Figure 4.
A multiagent simulation framework for P2P energy trading (DNO/DSO: Distribution network/system operator; TSO: Transmission system operator).

signals) with the coordinator (in centralised or distributed market designs) or with other participants (in decentralised market designs).

The upper block of **Figure 4** shows the external parties. Multiple P2P energy trading communities can trade and share energy with each other, creating a hierarchical P2P energy trading structure [15]. The P2P energy trading community needs to trade and share energy with the energy supplier (also called retailer) to balance the electricity surplus or deficit. Furthermore, the P2P energy trading community could provide various ancillary services, such as voltage support, congestion management and frequency support, for bulk power system utilities, such as distribution network operators (DNOs), distribution system operators (DSOs) and transmission system operators (TSOs) (**Box 5**) [16].

A day-ahead P2P energy trading among prosumers adopting the MMR mechanism is presented as an instance for demonstrating the multiagent simulation. For this instance, the framework in **Figure 4** is simplified as that in **Figure 5**.

As shown in **Figure 5**, using the MMR mechanism, the P2P energy trading coordinator decides the internal P2P energy trading prices, based on the energy bids provided by the prosumers and the import / export prices issued by the energy supplier. Four models are used to simulate the whole process of P2P energy trading, which are detailed as below.

1. The P2P energy trading coordinator agent and the MMR pricing model

The agent representing the P2P energy trading coordinator takes the inputs from the prosumer agents and energy supplier agent and generates the corresponding output, with the core being the MMR pricing model linking the inputs and outputs, as illustrated in **Figure 6**.

In **Figure 6**, $\boldsymbol{I}_{CA}$ is the set of inputs for the coordinator agent, where $\boldsymbol{p}^{SA} = \{p_t^{im}|t\in\boldsymbol{T}\}\cup\{p_t^{ex}|t\in\boldsymbol{T}\}$ is the set of electricity prices issued by the supplier agent. p_t^{im} represents the price of buying electricity from the supplier (i.e. import price). p_t^{ex} represents the prices of selling electricity to the supplier (i.e. export price). $\boldsymbol{T}$ is the set of time slots considered and t is the index of a time slot. $\boldsymbol{b}^{PA} = \left\{b_t^{PA}|t\in\boldsymbol{T}\right\}$ is the set of 'energy tenders' submitted by the prosumer agents, indicating the amount of energy to be sold (if negative) or bought (if positive). The set $\boldsymbol{O}_{CA}$ is the set of outputs for the coordinator agent. $\boldsymbol{p}^{P2P} = \{p_t^b|t\in\boldsymbol{T}\}\cup\{p_t^s|t\in\boldsymbol{T}\}$ is the prices for prosumers to conduct P2P energy trading with each other. p_t^b is the buying price and p_t^s is the selling price. $\boldsymbol{e}^{CA} = \{e_t^{CA}|t\in\boldsymbol{T}\}$ is the amount of electricity that the coordinator agent buys from / sells to the supplier agent (positive values for buying and negative for selling) for balancing the supply and demand of the whole P2P energy trading community.

Based on the 'energy tenders' from the prosumer agents, the P2P energy trading coordinator agent calculates the amount of energy that it needs to buy from / sell to the supplier agent to address the supply and demand imbalance within the P2P energy trading community:

$$e_t^{CA} = \sum_{i=1}^{N} b_{i,t}^{PA} \tag{6}$$

where N is the total number of prosumers in the P2P energy trading community and i is the index of a prosumer.

The P2P energy trading prices in the MMR mechanism, $\boldsymbol{p}^{P2P}$, are taken as the mean of grid import and export prices, with some modification terms when there is supply and demand imbalance in the P2P energy trading community, which needs to be balanced by the external supplier agent. Specific formulas are presented in **Appendix A**.

2. The prosumer agent and the decision-making model

The prosumer agent optimises the schedules of its flexible devices and the energy tender submitted to the P2P energy trading coordinator agent, in order to maximise its own economic benefits, as illustrated in **Figure 7**. Note that a prosumer agent can include models of distributed generators, energy storage and various electric loads, so being a general model that can describe other parties such as community energy storage and consumers as shown in **Figure 4**.

In **Figure** 7, $\boldsymbol{I}_{PA}$ is the set of inputs for the prosumer agent, which includes the P2P energy trading prices $\boldsymbol{p}^{P2P}$ issued by the P2P energy trading coordinator agent. The set $\boldsymbol{O}_{PA}$ is the set of outputs of the prosumer agent, which includes the 'energy bids' $\boldsymbol{b}^{PA}$ of the prosumer. The prosumer agent runs an optimisation to decide its optimal operation schedule and energy bids. The formulation of the optimisation problem is presented in **Appendix B**.

3. The energy supplier agent

The energy supplier is considered to act in a passive way, just to balance the electricity imbalances in the P2P energy trading community. The P2P energy trading coordinator imports/exports electricity from/to the energy supplier at the prices published by the energy suppliers. Flexible pricing schemes can be included, such as time-of-use (TOU) pricing, critical peak pricing (CPP), real-time pricing (RTP) and incline block pricing (IBP).

4. The implementation process and model

The implementation model specifies how the agents interact with each other, especially between the P2P energy trading coordinator agent and prosumer agents. The decisions of the coordinator agent and prosumer agents depend on each other, so iterations are used to describe their interaction. The iteration process can be fully implemented in practice, or used to model the gaming process in the agents' mind when they make decisions.
The convergence of iteration is an issue. In some designs, game-theoretic analysis is used to prove the existence of equilibria, where the iteration can converge, such as in [4]. However, game-theoretic methods usually assume models with specific mathematical forms, which might not be the case in practice. To address this issue, some heuristic methods, such as the 'Step Length Control' and 'Learning Process Involvement' proposed in [17], can be used to facilitate convergence. If convergence still cannot be achieved, the iteration process can be set to end after a pre-set number of iterations.
With the above formulation, P2P energy trading adopting the MMR mechanism can be simulated. A numerical example is given in Section 4.

Box 5.
Mathematical models for simulating P2P energy trading adopting MMR mechanism.

3.2 Evaluation index hierarchy

With the outcome of P2P energy trading obtained from the multiagent simulation or measured from real implementation cases, an evaluation system is needed for comprehensively assessing the performance of conducting P2P energy trading.

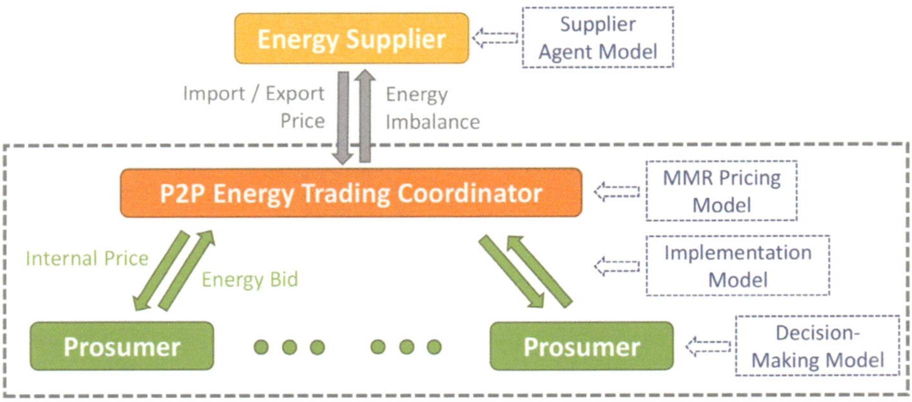

Figure 5.
The multiagent simulation framework for P2P energy trading adopting MMR mechanism.

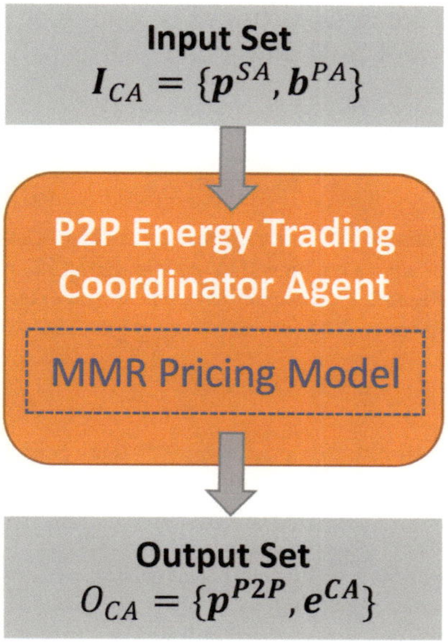

Figure 6.
The P2P energy trading coordinator agent.

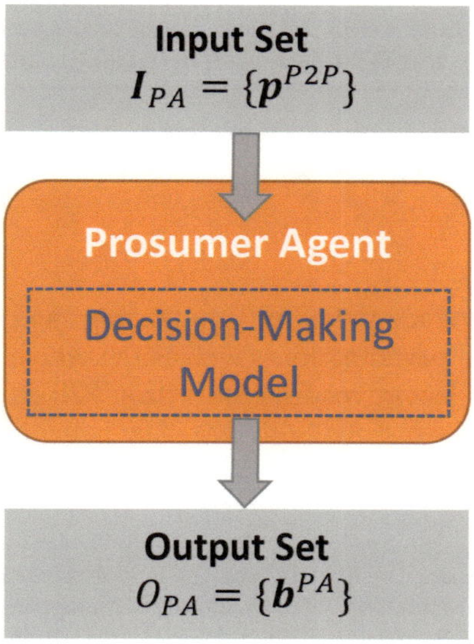

Figure 7.
The prosumer agent.

The impact of P2P energy trading can include a wide range of aspects – for example, a multi-dimension conceptual framework has been proposed to analyse blockchain-based P2P microgrids from technological, economic, social, environmental and institutional dimensions [17]. In this chapter, we will establish a general index hierarchy with some key quantitative technical and economic indexes for assessing the performance of P2P energy trading.

3.2.1 Technical indexes

From the perspective of microgrid/LES operators, one important motivation of conducting P2P energy trading is to better coordinate and integrate a high penetration of DERs. The potential technical capability of P2P energy trading is mainly in the improvement of local power and energy balance, which can be quantified in various ways.

Local Power Balance Index

The local power balance facilitated by P2P energy trading can help release circuit congestion, reduce circuit losses and improve the circuit utilisation within the microgrid/LES, and defer network reinforcement for both the microgrid/LES and upper-level power networks. The local power balance can be quantified by the peak-to-average ratio (PAR) within the microgrid/LES, as calculated by

$$I_{LPB} = \frac{P_M^{max}}{\overline{P_M}} \tag{7}$$

where I_{LPB} is the value of Local Power Balance Index; P_M^{max} is the peak active power of the microgrid/LES; $\overline{P_M}$ is average active power of the microgrid/LES throughout the time horizon considered (usually one day).

Local Energy Balance Index

Local energy balance indicates the extent to which local generation/demand can be consumed/ satisfied within the microgrid/LES, also reflecting the independency of the microgrid/LES. The local energy balance can be quantified by

$$I_{LEB} = \frac{\sum_{t=1}^{T}\left(P_t^{Im} + P_t^{Ex}\right)}{\sum_{t=1}^{T}\left(P_t^{D} + P_t^{G}\right)} \tag{8}$$

where I_{LEB} is the value of Local Energy Balance Index; P_t^{Im} and P_t^{Ex} are the total import and export power of the microgrid/LES at the time step t; P_t^D and P_t^G are the total demand and local generation of the microgrid/LES at the time step t; T is the total number of time steps considered (usually for one day).

Box 6.
Technical indexes for assessing P2P energy trading.

3.2.2 Economic indexes

The economic performance of P2P energy trading can be measured from the perspectives of individual participants as well as the whole microgrid/LES. Economic benefits are direct incentives for conducting P2P energy trading, especially for individual participants.

Coalition Stability Index

For each individual participant, the necessary condition for it to stay in the P2P energy trading community, rather than directly trade with the conventional energy supplier, is that its economic benefit obtained through P2P energy trading is no lower than those with the energy supplier. Therefore, the coalition stability of the P2P energy trading community can be quantified by the percentage of participants who have higher economic benefits in P2P energy trading than those with the energy supplier, i.e.

$$I_{CS} = \frac{N_H}{N_{P2P}} \tag{9}$$

where I_{CS} is the value of Coalition Stability Index; N_H is the number of participants who have higher economic benefits in P2P energy trading than those with the energy supplier; N_{P2P} is the total number of the participants in the P2P energy trading community.

Total Benefits Index

The total economic benefits of all the participants brought by P2P energy trading, with the benefits with the energy supplier as a reference value, are assessed by

$$I_{TB} = B_M^{P2P} - B_M^{ES} \tag{10}$$

where I_{TB} is the value of Total Benefits Index; B_M^{P2P} is the total economic benefits of all the participants in the P2P energy trading community; B_M^{ES} is the total economic benefits of all the participants if trading with the energy supplier otherwise. The economic benefits are measured by net electricity costs.

Benefits Allocation Index

Another index is needed for reflecting how the total benefits brought by P2P energy trading are allocated to each participant, which is defined as follows:

$$I_{BA} = \sum_{i=1}^{N_{P2P}} \sum_{j=1}^{N_{P2P}} \left| C_i - C_j \right| \tag{11}$$

where I_{BA} is the value of Benefits Allocation Index, and C represents the net cost of a P2P energy trading participant (positive for cost and negative for income) throughout the time horizon considered. It is seen that (11) calculates the summation of the cost difference between any two participants in P2P energy trading.

Box 7.
Economic indexes for assessing P2P energy trading.

3.2.3 Index normalisation and index hierarchy establishment

Based on the individual technical and economic indexes defined in the previous subsections, an index hierarchy can be established to assess the overall performance of P2P energy trading. To compare and synthesise different indexes, the indexes need to be normalised with the following features: (i) the index value being a number ranging from 0 to 1, and (ii) the index value '0' representing 'the worst' and '1' representing 'the best'. Specific normalisation of the indexes presented in the previous subsections is given in the following box. The technical and economic indexes presented in **Boxes 6** and 7 are normalised as follows (**Table 1**).

After normalisation, the indexes can be synthesised hierarchically by calculating the weighted sum, as presented below and illustrated in **Figure 8** [19]:

$$I_{TPI} = \sum\nolimits_{i}^{N_{TI}} \alpha_i \cdot I_i^{TI} \text{ where } \sum\nolimits_{i}^{N_{TI}} \alpha_i = 1, \alpha_i \geq 0 \tag{12}$$

Category	Index name	Original index	Normalised index	Notes
Technical	Local Power Balance	$I_{LPB} = \frac{P_M^{max}}{P_M}$	$I_{LPB}^* = 1 - \frac{I_{LPB}}{\sigma}$	σ: Maximum peak-average ratio from experience
	Local Energy Balance	$I_{LEB} = \frac{\sum_{t=1}^{T}\left(P_t^{Im} + P_t^{Ex}\right)}{\sum_{t=1}^{T}\left(P_t^{D} + P_t^{G}\right)}$	$I_{LEB}^* = 1 - I_{LEB}$	
Economic	Coalition Stability	$I_{CS} = \frac{N_H}{N_{P2P}}$	$I_{CS}^* = I_{CS}$	
	Total Benefits	$I_{TB} = B_M^{P2P} - B_M^{ES}$	$I_{TB}^* = \frac{I_{TB}}{B_M^{max} - B_M^{ES}}$	B_M^{max}: Maximum potential benefits that could be obtained (calculated through centralised optimisation [18])
	Benefits Allocation	$I_{BA} = \sum_{i=1}^{N_{P2P}} \sum_{j=1}^{N_{P2P}} \left\| C_i - C_j \right\|$	$I_{BA}^* = 1 - \frac{I_{BA}}{2N_{P2P}^2 \cdot \max_{i \in \{1,2...N\}} C_i}$	

Table 1.
Normalisation of evaluation indexes for assessing P2P energy trading.

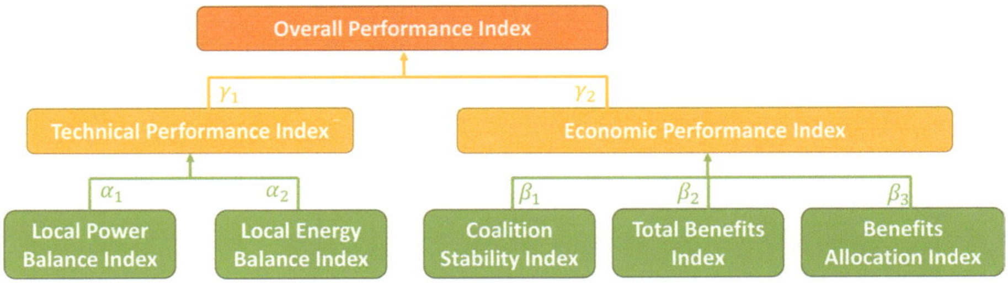

Figure 8.
Evaluation index hierarchy for assessing P2P energy trading.

$$I_{EPI} = \sum\nolimits_{i}^{N_{EI}} \beta_i \cdot I_i^{EI} \text{ where } \sum\nolimits_{i}^{N_{EI}} \beta_i = 1, \beta_i \geq 0 \tag{13}$$

$$I_{OPI} = \sum\nolimits_{i}^{N_{sub}} \gamma_i \cdot I_i^{sub} \text{ where } \sum\nolimits_{i}^{N_{sub}} \gamma_i = 1, \gamma_i \geq 0 \tag{14}$$

where I_{TPI}, I_{EPI} and I_{OPI} represent the values of Technical Performance Index, Economic Performance Index, and Overall Performance Index; α_i, β_i and γ_i are the weighting coefficients; I_i^{TI} represents the i-th technical index (such as the local power/energy balance index), and N_{TI} is the total number of technical indexes; I_i^{EI} represents the i-th economic index (such as Coalition Stability Index, Total Benefits Index and Benefits Allocation Index), and N_{EI} is the total number of economic indexes; I_i^{sub} represents the i-th index below the 'Overall Performance Index' level (such as Technical Performance Index and Economic Performance Index), and N_{sub} is the total number of these indexes.

It is worth noting that the indexes and hierarchy presented are just examples, and the methodology behind these examples is general and scalable. Firstly, more/different indexes for assessing technical or economic performance can be defined. Furthermore, other aspects beyond technical and economic performance, such as social, environmental and institutional dimensions, can also be considered to be assessed. More complicated methods can be used to synthesise the indexes at different levels, such as using the Analytic Hierarchy Process (AHP) method to decide the values of weighting coefficients.

4. Case study

A case study using multiagent simulation and the evaluation index hierarchy described in Sections 2 and 3 demonstrates P2P energy trading and its performance. A residential community consisting of 20 houses in Great Britain (GB) was considered. P2P energy trading was conducted between the customers within the community, adopting the MMR mechanism. The simulation was conducted in MATLAB and the optimisation was solved using CPLEX solvers.

4.1 Case design and parameters

For the 20 houses in the community, it was assumed that half of the houses had a 2 kW PV system for each house but did not have an electric vehicle, while each of the other half owned an electric vehicle but did not have a PV system. The CREST model [20], which is based on realistic GB statistics, was used to generate the electrical and usage patterns of the domestic appliances, demand and PV generation profiles of each house. The parameters and travel behaviour of the electric vehicles were generated from a database reflecting realistic use in GB [21].

The customers were assumed to conduct P2P energy trading under the MMR mechanism. They were assumed to do the trading day-ahead at the time resolution of one hour, i.e., making energy trading agreements for each hour of the next day one day in advance. For P2P energy trading, the 'Learning Process Involvement' technique was adopted as the implementation mechanism [19], with the learning rate selected as 0.5 and the maximum iteration number being 300. The import and export electricity prices provided by the external energy supplier were assumed to be 14.57 pence/kWh and 5.03 pence/kWh, which are typical values in GB.

Two reference cases (i.e., 'P2G' and 'Global Optimum') were simulated and compared with the P2P energy trading case (the 'P2P' case). The 'P2G' ('Peer to Grid') is the current arrangement in GB, where customers separately trade with the energy supplier at the import/export prices. In the 'Global Optimum' case, all the customers were assumed to be fully controlled by a centralised entity, which minimised the electricity cost of the whole community. The 'Global Optimum' case is to provide a reference to examine the capability of P2P energy trading in tapping the economic potential in the community.

When calculating the values of the evaluation indexes, equal coefficients were taken at all the levels.

4.2 Simulation and evaluation results – technical performance

The resulting daily demand profiles of the whole community in the three cases are presented in **Figure 9**, also with the community generation profiles and external/internal electricity prices plotted. The index values regarding the technical performance are shown in **Table 2**.

Figure 9(a) shows that the surplus PV power generation of some customers in the community was not consumed by other customers in the community due to the lack of coordination among customers when the customers separately traded with the external energy supplier. Moreover, the peak power of the community of the day was as high as 38.45 kW. By contrast, **Figure 9(b)** shows that, with the global optimisation, the local energy balance was able to reach a very high level, and the peak net power was reduced significantly to 10.14 kW.

Figure 9(c) shows the results with P2P energy trading. Compared to the 'P2G' case, much flexible demand was shifted to around noon to utilise the surplus PV power generation, incentivised by the lower internal P2P buying prices during those periods. This resulted in a much higher level of local energy balance as well as a higher level of local power balance with the peak net power of the community in the day being just 9.20 kW. Compared to the 'Global Optimum' case, the local energy balance level of the 'P2P' case was slightly lower expectedly, but the peak net power was also slightly lower (indicating an even better local power balance than the 'Global Optimum' case). This is because the price signals with the unit of £/MWh just incentivise local energy balance but do not incentivise local power balance directly. Therefore, additional measures, such as adding a 'capacity charge' element (i.e., with the unit being £/MW), can be taken (in the traditional retail market or in the P2P energy trading market) to further incentivise the local power balance.

Table 2 shows the quantified performance through the index values. Consistent with the results shown in **Figure 9**, the values of Local Power Balance Index, Local Energy Balance Index and Technical Performance Index in the 'P2P' case are 220.8%, 67.6% and 106.6% higher than those in the 'P2G' case. On the other hand, the value of Local Energy Balance Index in the 'P2P' case is just 5.5% lower than that

Figure 9.
The community demand and generation profiles with the external/internal electricity prices in the three cases. (a) the 'P2G' case. (b) the 'global optimum' case. (c) the 'P2P' case.

Cases	Local power balance index	Local energy balance index	Technical performance index
P2G	0.1591	0.4657	0.3124
P2P	0.5104	0.7805	0.6454
Global Optimum	0.3189	0.8260	0.5724

Table 2.
Indexes values of the technical performance in the three cases.

of the 'Global Optimum' case, showing that P2P energy trading under the MMR mechanism had got very close to the global optimum in this respect.

4.3 Simulation and evaluation results – economic performance

The total net cost of the community is illustrated in **Figure 10**, showing that P2P energy trading with the MMR mechanism significantly reduced the total net cost of the community compared to the 'P2G' case, being very close to the global optimum value.

Figure 11 shows that the daily net cost of any customer in the 'P2P' case was lower than that in the 'P2G' case, indicating that P2P energy trading benefited every customer to some extent and no customer would have the incentive to escape from the P2P energy trading coalition.

The index values regarding the economic performance are calculated as presented in **Table 3**. Note that the index values regarding the cost distribution were not calculated for the 'Global Optimum' case, since how the total cost of the community is allocated was not specified for the 'Global Optimum' case.

The numbers in **Table 3** are discussed as follows. Consistent with the results in **Figures 10** and **11**, the value of Total Benefits Index of the 'P2P' case is significantly higher than that of the 'P2G' case, being only 8.5% lower than that of the 'Global Optimum' case. No customer would be worse off in the 'P2P' case, so the value of Coalition Stability Index takes the maximum value of 1.0000. By contrast, the value

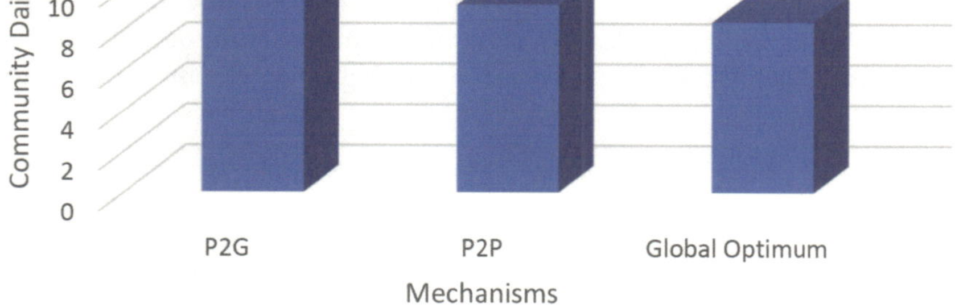

Figure 10.
The daily net cost to the community in the three cases.

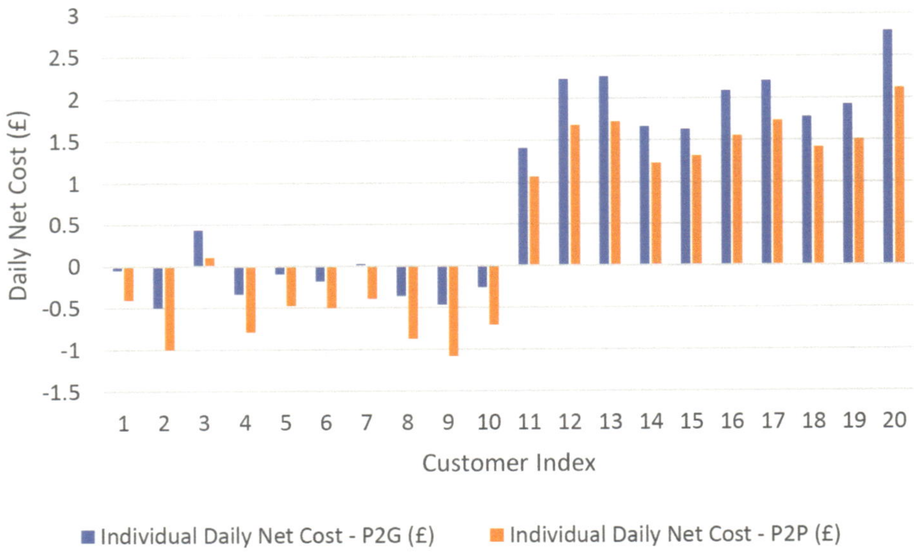

Figure 11.
The cost distribution in the community in the 'P2G' and 'P2P' cases.

Cases	Coalition stability index	Total benefits index	Benefits allocation index	Economic performance index
P2G	1.0000	0.0000	0.7738	0.5854
P2P	1.0000	0.9150	0.7060	0.8649
Global Optimum	N/A	1.0000	N/A	N/A

Table 3.
Indexes values of the economic performance in the three cases.

of Benefits Allocation Index in the 'P2P' case is slightly lower than that of the 'P2G' case, showing that P2P energy trading with the MMR mechanism did not improve the equity level in the community in terms of electricity cost. In summary, the value of Economic Performance Index of the 'P2P' case is 27.95% higher than that of the 'P2G' case, showing clear economic benefits brought by P2P energy trading.

4.4 Simulation and evaluation results – overall performance

The performance of the three cases is summarised in **Table 4**. It is shown that P2P energy trading has a significantly better performance from both the technical and economic perspectives, compared to the existing arrangement where customers

Cases	Technical performance index	Economic performance index	Overall performance index
P2G	0.3124	0.5854	0.4489
P2P	0.6454	0.8649	0.7552
Global Optimum	0.5724	N/A	N/A

Table 4.
Comparison of the overall performance of the three cases.

separately trade with the energy supplier, showing the great potential of P2P energy trading.

5. Summary

P2P energy trading is an innovative approach for managing increasing numbers of DERs in microgrids or local energy systems. In P2P energy trading, prosumers and consumers directly trade and exchange power and energy with each other. With proper design, P2P energy trading can create a triple win situation for customers, microgrids/local energy systems and wider bulker energy systems.

P2P energy trading is an emerging area with rapidly increasing academic research and industrial practice in many places of the world. P2P energy trading can be conducted across multiple temporal scales (e.g., settlement markets and forward markets with various time resolutions) and multiple spatial scales (i.e., within/between/beyond microgrids/LES). The development of P2P energy trading can be categorised into five key aspects, that is, i) market design, ii) trading platforms, iii) power & ICT infrastructure, iv) law, regulation & policy, and v) social science perspective.

Multiagent simulation and an evaluation index hierarchy were proposed to simulate and evaluate various P2P energy trading mechanisms. These techniques are useful for both academic study and industrial practice. In the multiagent simulation, relevant parties involved in P2P energy trading, such as market coordinators, energy suppliers and prosumers, are modelled as agents, with their internal behaviour models and interaction with other parties described. In the evaluation index hierarchy, quantitative indexes are defined to assess different aspects of P2P energy trading (e.g., technical and economic aspects), and indexes are synthesised to reflect higher-level performance. It is worth noting that the multiagent simulation framework and evaluation index hierarchy described are actually open systems, which are easy to extend and adjust for further development or specific applications.

Finally, P2P energy trading in a residential community in the context of GB was demonstrated as a case study. The houses had PV panels, electric vehicles and other typical appliances, trading and sharing electricity with each other in a day-ahead manner. Simulation results showed that the adopted P2P energy trading mechanism significantly improved the technical and economic performance of the community. The case study also showcased the application of a multiagent simulation framework and the effectiveness of the evaluation index hierarchy.

Appendix A – Pricing formulas in the Mid-Market Rate mechanism

The Mid-Market Rate mechanism makes the P2P energy trading prices using the following formulas:

$$p_t^b = \begin{cases} \dfrac{p_t^{mean} \cdot G_t + p_t^{im}(D_t - G_t)}{D_t} D_t > G_t \\ p_t^{mean} D_t \leq G_t \end{cases} \forall t \in \boldsymbol{T}, \tag{A1}$$

$$p_t^s = \begin{cases} p_t^{mean} D_t \geq G_t \\ \dfrac{p_t^{mean} \cdot D_t + p_t^{ex}(G_t - D_t)}{G_t} D_t < G_t \end{cases} \forall t \in \boldsymbol{T}, \tag{A2}$$

where

$$p_t^{mean} = \frac{(p_t^{im} + p_t^{ex})}{2} \forall t \in \boldsymbol{T}, \tag{A3}$$

$$D_t = \sum_{i=1}^{N} \max\left(b_{i,t}^{PA}, 0\right) \forall t \in \boldsymbol{T}, \tag{A4}$$

$$G_t = \left| \sum_{i=1}^{N} \min\left(b_{i,t}^{PA}, 0\right) \right| \forall t \in \boldsymbol{T}. \tag{A5}$$

In (A1) and (A2), p_t^b and p_t^s represent the prices at which the prosumers buy and sell electricity in the P2P energy trading community. p_t^{mean} is the average of the import price p_t^{im} and export price p_t^{ex} issued by the supplier agent, as calculated in (A3).D_t and G_t are the total electricity demand and supply in the P2P energy trading community, as specified by (A4) and (A5) respectively. $b_{i,t}^{PA}$ is the energy tender submitted by the prosumer agent i, indicating the amount of energy to be sold (if negative) or bought (if positive).

Appendix B – Optimal scheduling and bidding of prosumer agents

The prosumer agent runs an optimisation to decide its optimal operation schedule and energy bids. If a prosumer with solar PV panels, an electric vehicle, an electric water heater with a tank (as a flexible load) and inflexible electric loads is considered for example, the decision-making model of this prosumer is as follows:

$$\min \sum_{t=1}^{T} p_t^{P2P} \cdot P_t^{net} \cdot \Delta t \tag{B1}$$

where

$$P_t^{net} = x_t^H + \left(\frac{x_t^{ch}}{\eta^{ch}} + x_t^{dis}\eta^{dis}\right) + P_t^{inflex} - P_t^{PV} \forall t \in \boldsymbol{T}, \tag{B2}$$

$$p_t^{P2P} = \begin{cases} p_t^b \; P_t^{net} > 0 \\ p_t^s \; P_t^{net} \leq 0 \end{cases} \forall t \in \boldsymbol{T}, \tag{B3}$$

subject to

$$\sum_{k=1}^{t} x_k^H \Delta t + \rho \cdot M \cdot c \cdot (\theta_{ini} - \underline{\theta}) \geq \sum_{k=1}^{t} Q_k \; \forall t \in \boldsymbol{T}, \tag{B4}$$

$$\sum_{k=1}^{t} x_k^H \Delta t \leq \rho \cdot M \cdot c \cdot (\overline{\theta} - \theta_{ini}) + \sum_{k=1}^{t} Q_k \; \forall t \in \boldsymbol{T}, \tag{B5}$$

$$Q_t = \rho \cdot m_t \cdot c \cdot (\theta_{set} - \theta_{cold}) \; \forall t \in \boldsymbol{T}, \tag{B6}$$

$$0 \leq x_k^H \leq P^H \; \forall t \in \boldsymbol{T}, \tag{B7}$$

$$SOC_t = SOC_{ini} + \frac{1}{C} \sum_{k=1}^{t} (x_t^{ch} + x_t^{dis}) \Delta t \; \forall t \in \boldsymbol{T}, \tag{B8}$$

$$SOC_{t_{in}} = SOC_{t_{out}} - \Delta SOC \ \forall t \in \boldsymbol{T}, \tag{B9}$$

$$\underline{SOC} \leq SOC_t \leq \overline{SOC} \ \forall t \in \boldsymbol{T}, \tag{B10}$$

$$SOC_T = SOC_{ini} \ \forall t \in \boldsymbol{T}, \tag{B11}$$

$$0 \leq x_t^{ch} \leq P^{ch}, P^{dis} \leq x_t^{dis} \leq 0 \ \forall t \in \boldsymbol{T}. \tag{B12}$$

(B1) is the objective function, showing that the prosumer agent tries to minimise its daily net electricity cost. P_t^{net} is the net power consumption at the time slot t, calculated by (B2) where x_t^H is the power consumption of the electric water heater; x_t^{ch} and x_t^{dis} are the charging (always positive) and discharging (always negative) power of the electric vehicle; η^{ch} and η^{dis} are charging and discharging efficiencies of the electric vehicle; P_t^{inflex} is the sum of the all the inflexible loads; P_t^{PV} is the power generation of the PV panels. Positive P_t^{net} means the prosumer has electricity deficit, so the prosumer agent needs to buy electricity from the P2P energy trading community, thus the P2P energy trading price being the buying price shown in (B3). By contrast, negative P_t^{net} means the prosumer has electricity surplus, so the prosumer agent needs to sell electricity to the P2P energy trading community, thus the P2P energy trading price being the selling price shown in (B3). Δt is the length of a time step.

(B4)–(B7) are the constraints regarding the electric water heater with a hot water tank. (B4) and (B5) ensure that the thermal energy stored in the hot water tank can satisfy the hot water demand at the same time does not exceed the capacity of the tank. ρ is unit conversion coefficient between 'J' and 'kWh'; M is the maximum amount of water the tank can store in the tank; c is the specific heat capacity of water; $\underline{\theta}$ and $\overline{\theta}$ are the lower and upper limits of the water temperature in the tank; θ_{ini} is the initial water temperature in the tank. (B6) calculates the heat energy needed at each time step, Q_t, because of the hot water use. θ_{set} is the setpoint of the water temperature, and θ_{cold} is the temperature of the cold water inlet. m_t is the amount of water consumed at each time step. (B7) specifies the range of electric power of the water heater for heating.

(B8)–(B12) are the constraints regarding the electric vehicle. (B8) describes the evolution of the state of charge (SOC) of the batteries in the electric vehicle. (B9) models the impact of travelling on the SOC, where $SOC_{t_{in}}$ and $SOC_{t_{out}}$ represent the SOC when the electric vehicle returns home and leaves from home for travelling respectively, and ΔSOC represents the amount of energy needed (measured by SOC) for travelling. Note that here only one travel is modelled – if more travels are made, similar equations can be added. (B10) specifies the upper bound $\overline{SOC}$ and lower bound $\underline{SOC}$ of SOC. (B11) requires the SOC at the end of the day, SOC_T, should be equal to that at the beginning of the day, SOC_{ini}, so that the schedule is sustainable for the future. (B12) specifies the ranges of charging and discharging power.

In summary, given the P2P energy trading prices received from the P2P energy trading coordinator agent, the prosumer will optimise its flexible load schedule to minimise its own net electricity cost. The resulting net power consumption/generation will be submitted to the coordinator agent for deciding the P2P energy trading prices, i.e., the set of energy bids being $\boldsymbol{b}^{PA} = \left\{ b_t^{PA} = P_t^{net} | t \in \boldsymbol{T} \right\}$.

Author details

Yue Zhou* and Jianzhong Wu
School of Engineering, Cardiff University, UK

*Address all correspondence to: zhouy68@cardiff.ac.uk

References

[1] C. Zhang, J. Wu, Y. Zhou, M. Cheng, and C. Long, "Peer-to-peer energy trading in a microgrid," *Applied Energy*, vol. 220, pp. 1-12, Jun. 2018.

[2] Y. Zhou, J. Wu, C. Long, and W. Ming, "State-of-the-art analysis and perspectives for peer-to-peer energy trading," *Engineering*, vol. 6, no. 7, pp. 739-753, Jul. 2020.

[3] T. Morstyn, A. Teytelboym, and M. D. McCulloc, "Bilateral contract networks for peer-to-peer energy trading," *IEEE Transactions on Smart Grid*, vol. 10, no. 2, pp. 2026-2035, 2019.

[4] N. Liu, X. Yu, C. Wang, C. Li, L. Ma, and J. Lei, "Energy-sharing model with price-based demand response for microgrids of peer-to-peer prosumers," *IEEE Transactions on Power Systems*, vol. 32, no. 5, pp. 3569–83, 2017.

[5] D. H. Nguyen, "Optimal solution analysis and decentralized mechanisms for peer-to-peer energy markets," *IEEE Transactions on Power Systems*, vol. 36, no. 2, pp. 1470-1481, Mar. 2021.

[6] W. Tushar, C. Yuen, H. Mohsenian-Rad, T. Saha, H. V. Poor, and K. L. Wood, "Transforming energy networks via peer-to-peer energy trading: the potential of game theoretic approaches," *IEEE Signal Process Magazine*, vol. 35, no. 4, pp. 90-111, 2018.

[7] C. Long, J. Wu, C. Zhang, L. Thomas, M. Cheng, and N. Jenkins, "Peer-to-peer energy trading in a community microgrid," in *Proceedings of IEEE PES General Meeting*, Chicago, IL, USA, pp. 1-5, 2017.

[8] M. Walport. "Distributed ledger technology: beyond block chain." Government Office for Science. 2016. https://assets.publishing.service.gov.uk/government/uploads/system/uploads/attachment_data/file/492972/gs-16-1-distributed-ledger-technology.pdf (accessed Apr. 13, 2021).

[9] E. Mengelkamp, J. Gärttner, K. Rock, S. Kessler, L. Orsini, and C. Weinhardt, "Designing microgrid energy markets: a case study: the Brooklyn microgrid," *Applied Energy*, vol. 210, pp. 870-880, Jan. 2018.

[10] O. Jogunola, A. Ikpehai, K. Anoh, B. Adebisi, M. Hammoudeh, H. Gacanin, and M. Hammoudeh, "Comparative analysis of P2P architectures for energy trading and sharing," *Energies*, vol. 11, no. 1, 2018.

[11] C. Zhang. Peer-to-peer energy trading in electrical distribution networks [PhD Dissertation]. Cardiff: Cardiff University; 2017.

[12] A. Singh, A. T. Strating, N. A. R. Herrera, D. Mahato, D. V. Keyson, and H. W. van Dijk, "Exploring peer-to-peer returns in off-grid renewable energy systems in rural India: an anthropological perspective on local energy sharing and trading," *Energy Research & Social Science*, vol. 46, pp. 194-213, Dec. 2018.

[13] A. Singh, A. T. Strating, N. A. R. Herrera, H. W. van Dijk, and D. V. Keyson, "Towards an ethnography of electrification in rural India: social relations and values in household energy exchanges," *Energy Research & Social Science*, vol. 30, pp. 103-115, Dec. 2017.

[14] F. Ecker, *H. Spada*, and U. J. J. Hahnel, "Independence without control: autarky outperforms autonomy benefits in the adoption of private energy storage systems," *Energy Policy*, vol. 122, pp. 214-228, 2018.

[15] L. Shi, Y. Zhou, C. Long, S. Abeysinghe, L. Cipcigan, and J. Wu, "A peer-to-peer energy trading hierarchy for microgrids in distribution

networks," *in Proceedings of International Conference on Applied Energy*, Västerås, Sweden, pp. 1-6, 2019.

[16] Y. Zhou, J. Wu, G. Song, and C. Long, "Framework design and optimal bidding strategy for ancillary service provision from a peer-to-peer energy trading community," *Applied Energy*, vol. 278, 115671, 2020.

[17] A. Ahl, M. Yarime, K. Tanaka, and D. Sagawa, "Review of blockchain-based distributed energy: Implications for institutional development," *Renewable and Sustainable Energy Reviews*, vol. 107, pp. 200-211, Jun. 2019.

[18] Y. Zhou, J. Wu, C. Long, M. Cheng, and C. Zhang, "Performance evaluation of peer-to-peer energy sharing models," in *Energy Procedia (World Engineers Summit – Applied Energy Symposium & Forum: Low Carbon Cities & Urban Energy Joint Conference, WES-CUE 2017)*, Singapore, pp. 1-6, 19-21 Jul. 2017.

[19] Y. Zhou, J. Wu, and C. Long, "Evaluation of peer-to-peer energy sharing mechanisms based on a multiagent simulation framework," *Applied Energy*, vol. 222, pp. 993-1022, Jul. 2018.

[20] E. McKenna, and M. Thomson, "High-resolution stochastic integrated thermal-electrical domestic demand model," *Applied Energy*, vol. 165, pp. 445-461, Jan. 2016.

[21] Y. Zhou, J. Wu, and C. Long, "Electric_Vehicle_Dataset (Great Britain).xlsx", in *Evaluating peer-to-peer energy sharing mechanisms for residential customers in present and future scenarios of Great Britain (Cardiff University Research Datasets)*, available from: http://doi.org/10.17035/d.2018.0046405003

Section 2

Local Energy Systems

Chapter 4

District Heating and Cooling Systems

Iván De la Cruz and Carlos E. Ugalde-Loo

Abstract

Decarbonisation of the energy sector is a crucial ambition towards meeting net-zero targets and achieving climate change mitigation. Heating and cooling accounts for over a third of UK greenhouse emissions and, thus, decarbonisation of this sector has attracted significant attention from a range of stakeholders, including energy system operators, manufacturers, research institutions and policy makers. Particularly, the role of district heating and cooling (DHC) systems will be critical, as these two energy vectors are central to our lives not only for comfort and daily activities, but also to facilitate productive workplaces and to run a variety of industrial processes. The optimal operation of DHC systems and the design of efficient strategies to produce heat and cold, store thermal energy, and meet heating and cooling demands, together with an increased integration of low carbon technologies and local renewable energy sources, are vital to reduce energy consumption and carbon emissions alike. This chapter reviews relevant aspects of DHC systems, their main elements, automatic control systems and optimal management.

Keywords: District heating and cooling systems, local energy systems, sustainable energy systems, renewable energy sources, thermal energy storage, heat transfer, heating demand, cooling demand, energy systems management

1. Introduction

From very early days, men and women learned to manipulate heat for their protection and survival. From the discovery of fire to the development of thermal systems in modern societies, the supply of thermal energy to meet heating and cooling demands has significantly improved, both in the efficiency of transport and the quality of the thermal energy available. As we face an unprecedented climate emergency threatening life in our planet, mitigation of global warming through the reduction of greenhouse gas emissions has become a crucial ambition. Due to the substantial amount of carbon emissions contributed by domestic and industrial heating, including space cooling, district heating and cooling (DHC) systems have recently attracted significant attention. With suitable operating regimes, inclusion of low carbon technologies and control strategies in place, a DHC system may be an integral solution for the reduction of energy consumption and carbon emissions while meeting local demand for heat and cold.

DHC systems use the infrastructure interconnecting dwellings, buildings or facilities within a city, district or neighbourhood to supply heating or cooling [1]. Using mainly water as the transport medium, heat and cold are distributed to meet demand for space heating and cooling, refrigeration, and water for domestic or industrial use.

A local energy system is an energy system led by a local authority or local organisations within a bounded geographical area (commonly a city) that aims to match local energy resources with the local energy demand [2, 3]. Therefore, a DHC system is considered a local energy system as it is limited to specific areas within a city, although it can extend to an entire municipality. In occasions a thermal system spans beyond municipal borders to incorporate thermal energy generation plants, such as Copenhagen's district heating system [3].

DHC systems supply a large share of the thermal energy demand in a local region. Therefore, considering them in regional plans for reaching net-zero targets is of importance. Although exploiting local energy resources could lead to DHC systems being part of energy islands [3, 4], they would be connected to national energy systems through electricity and gas grids. Thus, coordinated actions of national and local plans are required for a successful transition incorporating renewable energy sources (RESs).

The main components of a DHC system are shown in **Figure 1**: heating or cooling energy sources, distribution network components, customer installations including thermal loads and thermal energy storage (TES) systems.

1.1 Evolution of district heating and cooling systems

The first district heating system dates back to 1877, in New York, where 14 customers were connected to a network supplied by a single boiler fuelled by coal [5]. Subsequent systems developed elsewhere worked on the same principle. In general, they consisted of steam traps, expansion joints, and insulated steam pipes

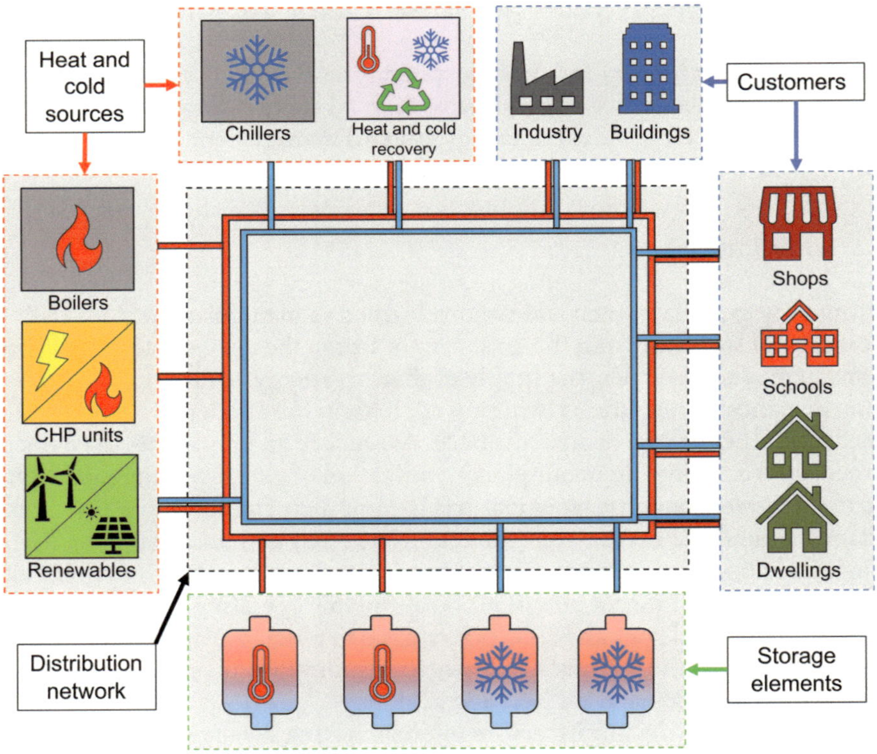

Figure 1.
Classification of components of a DHC system.

(supply and return), housed within underground concrete ducts transporting steam at high temperature and high pressure. Steam heat storage plant and combined heat and power (CHP) generation units driven by coal or oil were also incorporated, but their use was not extensive [6]. Steam district heating systems were highly inefficient due to the large heat losses resulting from operation at high temperatures. In addition, these systems were hazardous due to the risk of explosion. Steam district heating systems were the 1st generation of district heating (1GDH) systems, which dominated the market until the 1930s.

The 2nd generation of district heating (2GDH) systems replaced steam with pressurised liquid water at temperatures above 100°C, making the systems safer while reducing heat losses. Water pipes and water-based heat storage tanks were used. Large shell-and-tubes heat exchangers and large and heavy valves were also incorporated to the distribution network. The number of CHP units also increased, further integrating heating and electricity networks, while decreasing costs of energy production and emissions at the same time.

Although 2GDH systems were dominant until the 1970s and a few 1GDH and 2GDH systems remain operational, the 3rd generation of district heating (3GDH) systems is predominant nowadays. The distinctive feature of a 3GDH system is the use of heat interface units to provide an individual supply temperature control. Liquid water is still employed in the primary circuit; however, as the operating temperatures are below 100°C, heat losses and insulation requirements for pipelines are reduced. Large gas, biomass and waste CHP units replaced oil- and coal-driven units. Moreover, a range of distributed heat sources such as small CHP units, gas and electric boilers, solar collectors and heat pumps were incorporated, thereby integrating RESs as heat sources.

The development of district cooling systems followed a similar approach as district heating systems. Introduced in the 19th century, the 1st generation of district cooling (1GDC) systems were predominant until the 1960s. They consisted of centralised condensers and decentralised evaporators, with a refrigerant used as the transport medium. A solution of ammonia and saltwater was often used as the refrigerant. The 2nd generation of district cooling (2GDC) systems used chilled water as the distribution medium and adopted large mechanical chillers to produce cold. The 3rd generation of district cooling (3GDC) systems is dominant these days and grew in popularity in the 1990s. With a diversified cold supply consisting of absorption and mechanical chillers, modern cooling systems take advantage of local natural water resources, such as lakes or rivers, to obtain free cooling.

1.2 Modern district heating and cooling networks: an insight to Paris' system

According to the United Nations Environment Program, Paris is considered one of the "champion cities" for district energy use in the world [7]. With both large district heating and district cooling networks currently in operation, Paris is aiming to reduce its carbon emissions by 75% by 2050.

District heating was introduced in Paris in 1927 through a 1GDH system configuration based on steam. With 33% owned by the city, the Paris Urban Heating Company (CPCU) is still the only company operating under a concession contract. It currently serves 16 towns in the metropolitan area, supplying heat to ~500,000 households using a network of 475 km. Although steam is used as the main distribution medium, the network also includes smart operation control systems and several hot water loops (as those used in a 3GDH configuration). Due to their high distribution efficiency, hot water loops are preferred for areas under development in the city suburbs. These, in turn, help reduce peak demand on the steam network.

CPCU uses a wide range of heat sources. Three waste incinerators constitute the primary heat production units, which are supported by two CHP plants, several gas and coal boilers, and five fuel oil boilers. There are plans for replacing gas and coal boilers with waste and biomass-fired power plants by 2030 [8]. Currently, 50% of the energy consumed by the district heating network comes from RESs and energy recovery. As a result, the Paris district heating system is an important example of heat generation and supply from renewable energy.

Paris also has the most extensive 3GDC system in Europe. The cooling system has been operated by Climespace since 1991. It has an 83 km underground cooling network, which supplies chilled water to over 700 customers, including hotels, shopping centres, offices, state buildings, museums, and theatres. The system employs ten cooling energy production sites to produce ~480 GWh of cooling per year and it also features four cooling storage facilities [9]. What makes this system highly economic, operationally efficient and innovative is the incorporation of free cooling into the energy production process. Water from the Seine River is used as a large natural source of chilled water for the distribution network, reducing both costs and the environmental impact.

Paris' district heating and district cooling networks are monitored and operated by specialised personnel in real time through control centres to guarantee a consistent supply of heating and cooling. Both networks are constantly being extended, incorporating additional delivery stations to reduce the city's carbon footprint.

1.3 Modern district heating and cooling systems

1.3.1 The 4th generation district heating and district cooling systems

Existing 3GDH and 3GDC systems and their technologies are evolving to new systems. The 4th generation of district heating (4GDH) systems and the 4th generation of district cooling (4GDC) systems integrate RESs to ensure efficient and high-performance operation [6]. Although 4GDH and 4GDC systems are designed as individual networks, exploiting synergies between them is required to use energy efficiently, especially for recovering waste heating and cooling energy.

The expansion of heating and cooling distribution networks by integrating local RESs and exploiting the potential for recovering waste thermal energy from a wide range of customers' applications brings substantial benefits. Combined with a smart management of the systems and of the customers' installations, the overall efficiency of the thermal grid may be improved and the difference between supply and return temperatures reduced with respect to the ambient temperature. This temperature reduction could in turn help minimising thermal losses. Intelligent management includes 24-hour weather forecasting to estimate thermal demand and the use of load shedding and load shifting techniques to regulate it. Building insulation for energy conservation and the minimisation of health risks posed by the Legionella bacteria by eliminating domestic hot water (DHW) stores enables the reduction of water supply temperatures for space heating, space cooling and DHW production.

Thermal losses in the distribution network are one of the main concerns for future developments. In addition to a reduction of temperature in supply and return lines, achieving low thermal losses requires:

- Reducing pipe dimensions.
- Improving the insulation of pipelines using twin pipes.

- Implementing 3-pipe configuration in the distribution network to maintain a supply temperature at the periphery, avoiding bypassing the return line [10].
- Implementing a continuous metering system with fast wireless readings, enabling intelligent control of the network.

Decarbonisation of thermal networks involves the use of RESs and other low-carbon heating and cooling sources, which include CHP plants and efficient boilers, waste incineration plants, chillers and heat pumps. For local energy systems, the integration of energy resources available in close proximity is desirable. For a thermal network, these include free cooling from nearby natural water reservoirs and geothermal energy from low-temperature geothermal heating plants with efficiently operated heat pumps. Solar thermal plants can also be integrated into the network when combined with large-scale seasonal TES systems. Waste heat and cold from local industrial activity and commercial buildings represent a significant thermal resource. To this end, central thermal stores are used to recover waste thermal energy for a later dispatch.

Important benefits are obtained when operating thermal networks with supply and return temperatures near the ambient temperature. The capacity of a thermal network to take heating (or cooling) energy from a source is enhanced by lowering its return temperature (or raising it for cooling) since heat transfer is proportional to the temperature difference between the source and the network. For this reason, steam or fuel-driven CHP plants enhance their electricity production by enabling a further expansion of steam or flue gases without affecting the total heat recovered. Heat pumps, chillers and TES units are also sensible to temperature. A higher coefficient of performance (CoP) in vapour compression heat pumps and chillers is achieved by reducing the temperature difference between the heat source and heat sink since less work is required from the compressor. Charging and discharging temperatures limit the energy storage capacity of a TES unit. Since the discharge temperature is the return temperature in the network, lowering this temperature in TES units storing heat (or raising it for cold stores) can increase storage capacity. The design of smart energy systems aims to cope with the fluctuating nature of RESs. To achieve this, electricity, thermal and gas networks must be combined and their operation coordinated at a local and national level, providing optimal solutions at an individual level and as part of an integrated network. Coupling devices linking energy vectors, such as heat pumps, CHP plants and chillers, and the use of energy storage components play a crucial role in the operation of a local thermal network.

The large-scale integration of RESs involves a radical technological change and a paradigm shift from single-purpose organisations implementing undifferentiated solutions to multi-purpose organisations implementing differentiated solutions. Three main challenges have been identified:

- Deciding whether to have district heating and district cooling systems or individual thermal solutions.
- Establishing a trade-off between developing local heat production against implementing energy conservation measures.
- Motivating the integration of local RESs.

To address these challenges, it is necessary to develop planning and design tools and methodologies based on geographical information systems. On the other hand, implementing long-term marginal cost tariff policies and enabling access to loans and consultancy services to support customers are also relevant measures.

Although the technologies and policies for 3rd generation systems are in an early transition towards the 4th generation, encouraging results in research projects have been obtained when implementing the measures discussed previously [11]. Fourth generation systems are expected to dominate modern thermal systems in the future.

1.3.2 The 5th generation of district heating and cooling

Until the 4th generation, district heating and district cooling networks were treated as individual loops with limited interaction between them supplying either heat or cold to customers. The systems were unidirectional, with water flowing in one direction through the distribution network to meet thermal demands. However, in the last few years, a novel configuration using bidirectional networks has emerged. The 5th generation of district heating and cooling (5GDHC) systems integrate district heating and district cooling networks into a single distribution network [12–14], delivering either heating or cooling to a single facility by inverting the direction of the water flow in the consumer terminals. These systems reduce supply and return temperatures even further by using decentralised heat pumps. Benefits from 5GDHC systems include the reduction of thermal losses in the distribution grid, improved TES capacity, and an increased utilisation of heating and cooling generation plants and waste heat and cold sources. **Figure 2** shows a schematic comparison between 3rd and 4th generation district heating and district cooling systems and a 5GDHC system.

The characteristics of a 5GDHC system make of it a powerful and versatile solution to decarbonise thermal networks. For instance, bidirectionality of the flow produces different operational mass flow rates for different subsections of the distribution network. This feature enables to exploit synergies between buildings with different thermal demands, recovering waste heat or cold from some premises to supply others within the same cluster of buildings. Circular use of the energy may be achieved, where consumers become prosumers supplying the network. This could enable peer-to-peer energy transaction schemes to reduce energy costs. Bidirectionality also facilitates the integration of RESs and local thermal energy sources into the network.

Deployment of 5GDHC systems is challenging. Adding decentralised heat pumps to allow reduced working temperatures requires a significant capital investment and sophisticated control systems since the heat pumps are connected to the electricity grid. Reduction in temperatures also requires higher mass flow rates, for

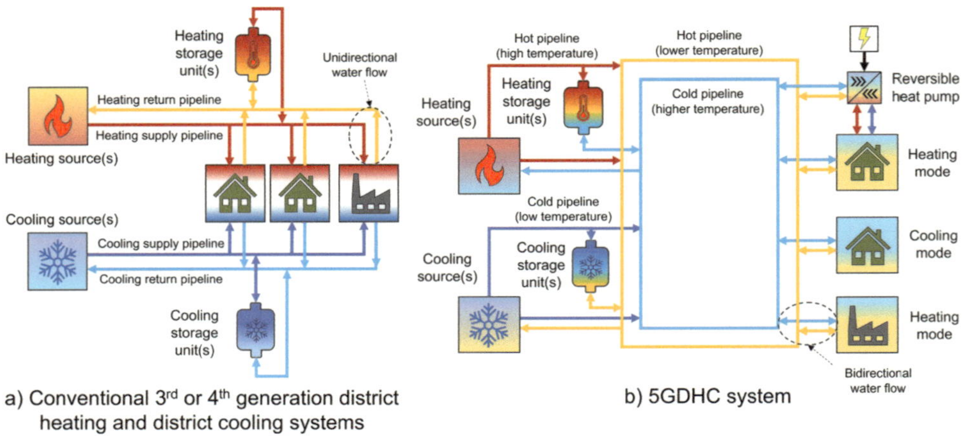

Figure 2.
Schematic comparison between 3rd or 4th generation district heating and district cooling systems with 5GDHC systems.

which extra pressure boost or larger diameter pipes would be required. Despite these challenges, there are 5GDHC systems in operation integrating RESs, including groundwater (e.g. ETH Campus Hönggerberg, Switzerland), seawater (e.g. Bergen University, Norway), or excess heat (e.g. Herleen, The Netherlands) [13, 14].

2. Thermal loads in energy transfer stations

Thermal energy flows in a DHC system can be considered as either supply or consumption flows. Supply flows are produced at the ends of the distribution network connected to the heat or cold sources. Conversely, consumption flows are generally associated with the customer ends or energy transfer stations (ETSs), which connect the customer's local installation to the distribution networks via heat exchangers or hydraulic separators.

Consumption flows are commonly referred to as thermal loads [15] and expressed in kilowatt [kW] or kilowatt-hour [kWh]. Thermal loads change with time and are specified as load profiles ranging from an annual, seasonal, monthly, weekly, daily, or even hourly basis [15, 16]. Weekly and daily thermal load profiles depend on the level of activity within the network which, in turn, depends on the energy consumption of each customer. Monthly and seasonal load profiles depend mainly on meteorological conditions. For instance, the demand for heating is high in winter, while cooling demand may dominate in summer. The definition of load profiles is critical for the optimal operation of DHC systems [17]. This will be briefly discussed in Section 6.

Within local customer installations, thermal loads are caused by three main energy consumption processes: space heating or space cooling, production of DHW, and industrial processes. Additionally, thermal losses in the distribution network represent a significant portion of the total thermal load.

2.1 Space heating and space cooling

Space heating and space cooling are used to condition the temperature inside a dwelling, building or facility to a uniform level which is different from the outdoors temperature. This may be achieved by placing radiators in each room or at specific locations inside a building. Alternatively, fan coil units distributed through a building regulate indoor temperature by varying the ventilation airflow temperature and mass flow rate [18]. Thermostats are included in key areas of a building when automatic systems are implemented to regulate temperature.

Lighting, the temperature of the water sources, or the insulation level affect the conditioning of indoor temperature within a building [19, 20]. The number of building occupants is also a major factor. For example, the energy consumption for space cooling in an office complex in the summer is higher during working hours than for the rest of the day. Meteorological conditions are particularly relevant for space heating and space cooling. Although these can lead to a higher heating or cooling consumption, they can also help to mitigate thermal losses due to the difference in indoor and outdoor temperatures. Solar radiation is especially beneficial for space heating during winter. Similarly, rainy or windy days lead to a lower cooling consumption in the summer.

2.2 Domestic hot water preparation

DHW refers to the water used for domestic purposes including cooking, sanitation and personal hygiene, where the final consuming elements are taps and showers. Thermal load due to DHW depends primarily on the specific temperature setpoint the water is heated to and the activity level of the customer.

In 3GDH and 4GDH systems, DHW preparation involves the use of water tanks and strict temperature control. To prevent the growth of Legionella, the bacteria causing Legionnaires' disease, temperature must be kept above 50°C [21]. In 5GDHC systems, this is challenging due to the low temperatures of operation. For this reason, heat pumps and electric heaters are used to locally raise temperature [22, 23], which increases the coupling between thermal and electrical networks.

2.3 Industrial thermal loads

Industrial loads can be attractive for DHC systems thanks to the diversity of thermal processes employed to manufacture and store products. The transport, electricity, pharmaceutics, and food sectors, and manufacturing processes such as machinery, textiles, chemicals, paper, metal, wood, rubber and plastic require heating and cooling [24, 25]. Although thermal loads in an industrial process may be affected by outdoor temperature and meteorological conditions, the determining factor to predict load profiles is the periodicity of activities within the process [24, 26].

Due to the economic and environmental benefits involved, the incorporation of industrial activity into DHC systems is increasing. In addition, the recovery of industrial waste heat and cold represents a major thermal energy resource that could be employed to meet the demands from other customers connected to the thermal network [25, 27]. Further discussion on this is provided in Section 3.

2.4 Thermal losses in the distribution network

Thermal losses modify the overall thermal load of a network. Several factors affect the amount of energy lost. The network's integrity is a major factor since cracks in distribution elements can leak water [6, 28]. It is thus essential to regularly monitor the network and to provide maintenance when required.

The amount of insulation and the geometry of pipes (i.e. diameter and length) influence the network's heat transfer to the surrounding ground [20, 28]. The reduction of operating temperatures in the latest generations of DHC systems aims to reduce thermal energy losses resulting from this heat transfer.

3. Heating and cooling sources

3.1 Boilers and burners

Depending on the heat source, boilers are classified as boilers supplied by fossil fuels, heat recovery boilers and electric boilers. Electric boilers (i.e. electric resistances immersed into water tanks) and small-scale fuel powered boilers are mainly used to produce hot water for space heating and DHW in dwellings, housing developments or offices complexes. Heat recovery and combustion boilers are extensively used in DHC systems and in industrial applications due to their wide range of capacities.

3.1.1 Fossil fuel fired boilers

Fossil fuel fired boilers consist of two separate flow channels to transfer heat from the fuel to water. Water is fed and circulated through the first channel. At the inlet ports of the second channel, fuel and air are fed and burnt to generate hot flue gases. These gases circulate through the second channel, transferring heat to the water at the points where both flow channels are linked.

Fossil fuel fired boilers may vary depending on the type of fuel used, the firing method, the field of application, the water circulation system, or the operating pressures [29]. However, they are classified mainly as hot water boilers and steam boilers. Steam boilers are more complex because of the separation process of the steam from liquid water. A schematic of a steam boiler is shown in **Figure 3**.

3.1.1.1 Economiser

An economiser consists of a heat exchanger where the feedwater temperature is initially raised, taking heat from the flue gases at the boiler exhaust. Flue gases at this point are at a lower temperature than in the furnace.

The economiser heat exchanger is found in both hot water and steam boilers. It is used to recover heat from the flue gases before they are released into the atmosphere, thereby increasing fuel utilisation.

3.1.1.2 Steam drum

Hot water boilers often feed water directly from the economiser to the tubes of the furnace, but in a steam boiler the preheated liquid water going out of the economiser can be passed to one or a set of steam drums. Within the steam drum, the fraction of water at liquid state is separated from the steam coming from the furnace, and it is fed back to the furnace through downcomers to be reheated.

3.1.1.3 Furnace

Most of the heat produced by fuel combustion is transferred to the water in the furnace, which can be configured in two different ways (see **Figure 4**). In fire-tube

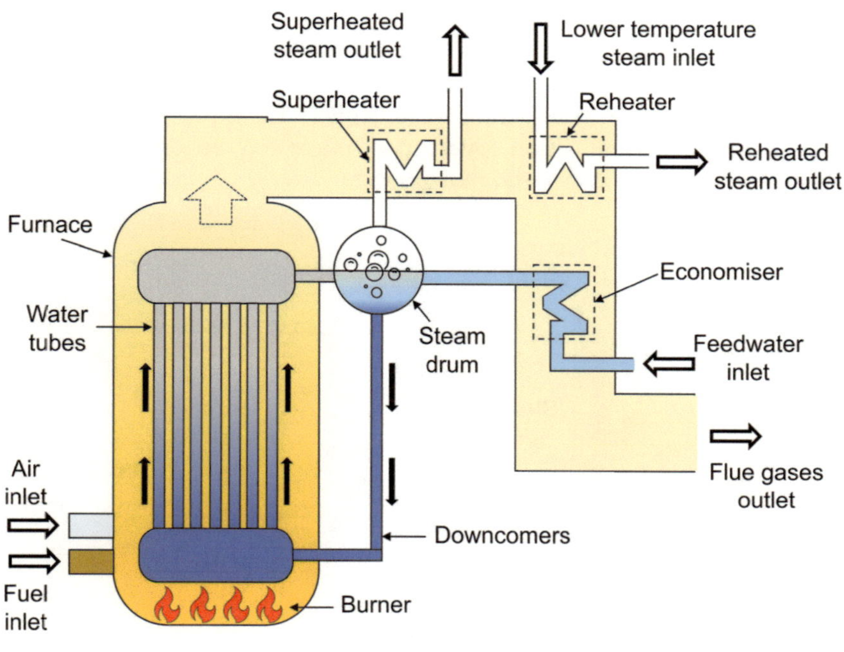

Figure 3.
General configuration of a steam boiler.

boilers (left), the flue gases flow through metal tubes placed within the water flow. Conversely, in water-tube boilers (right), water flows through an arrangement of vertical tubes between the flow of flue gases.

3.1.1.4 Superheater and reheater

These are exclusive to steam boilers. Heat exchangers are placed through the exhaust conduit to increase the steam temperature in several phases. Thus, they increase fuel utilisation by drawing heat from the flue gases. These heat exchangers can be used to superheat the steam exiting the furnace, or to reheat low temperature steam that has been cooled down by a process. In some configurations, superheaters may have their own combustion system to support the steam heating, which requires extra fuel consumption.

3.1.2 Heat recovery boilers

These work under the same principle of fossil fuel boilers. However, heat recovery boilers use the heat generated by burning the by-products of industrial processes instead of fossil fuel. For example, Kraft boilers used in the pulp and paper industry exploit the burning process of black liquor—a by-product of the pulp-making process [29, 30].

3.1.3 Waste heat recovery boilers

The waste heat recovery boiler is different from other boilers as it does not have a furnace. This boiler is powered by the heat generated from processes that would be wasted in normal conditions. This boiler is a fundamental part of CHP plants [31].

3.1.4 Operation and control of boilers

Boiler operation is monitored by a set of temperature, pressure and flow rate sensors. On–off and proportional-integral controllers are implemented to regulate internal variables. For combustion boilers, controllers and sensors are implemented for flow channels of both water and flue gases.

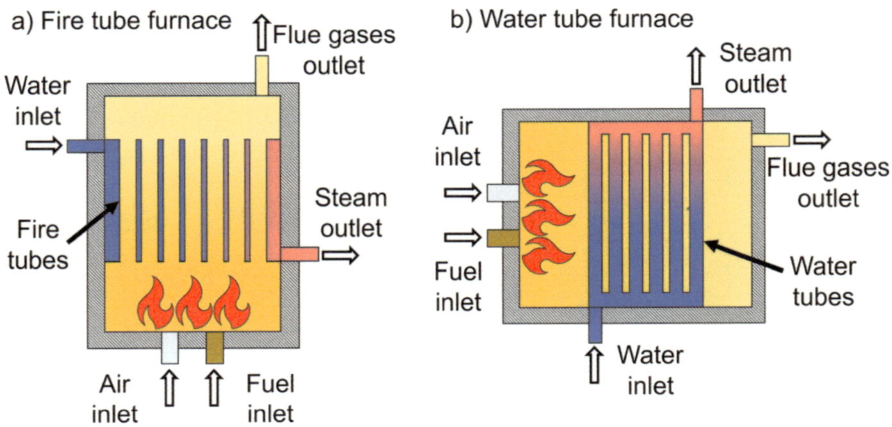

Figure 4.
Fire-tube and water-tube furnace configurations.

The fuel and air mass flow rates are responsible for the heat generation within the flue gases flow channels and thus the temperature and pressure in the furnace. Two control systems are implemented to regulate this operation: the burner management system and the firing rate control [32]. The burner management system supervises and limits the burner operation for safety, while the firing rate control regulates the fuel and air flows. The air inflow rate is regulated to achieve a desired air-to-fuel ratio, which guarantees the fuel is completely burnt while controlling sulphur and nitrogen oxides emissions [29].

The water flow channel is operated by the feedwater flow control system and, depending on the type of boiler, by the steam pressure or the fluid temperature maintenance system. The feedwater flow system is driven by the boiler feed pumps and, in some cases, by additional circulating pumps, which control the feedwater flow and fluid level. In hot water boilers, the outlet water temperature is maintained at a specific setpoint by varying the feedwater flow. However, the control of steam boilers is more complex. While the feedwater flow system keeps a constant water level in the drum, the pressure in the steam conduit is maintained by regulating the amount of fuel burnt and, thus, the steam flow produced.

Effective monitoring and control of the boiler operation are required to achieve an optimal performance, reduce losses, and draw the highest efficiency. The conventional definition of a fuel fired boiler's efficiency η_{boiler} is given by

$$\eta_{boiler} = \frac{Q_{heat}}{Q_{fuel}} = \frac{\left(Q_{fuel} - \text{losses}\right)}{Q_{fuel}} \tag{1}$$

where Q_{heat} is the total heat absorbed by water or steam considering losses, and Q_{fuel} is the heat added by the fuel—which can be approximated by the fuel's lower heating value or higher heating value.

3.2 Chillers

Chillers use the thermodynamic properties of refrigerant fluids to extract heat from water or air streams to produce cooling. Chillers are classified according to their refrigeration cycle.

3.2.1 Mechanical vapour compression chillers

Mechanical vapour compression chillers use mechanical power to drive a compressor. Although the compressor can obtain power from an engine or a turbine, the most popular power source is electricity. **Figure 5** shows the refrigeration cycle of a vapour compression chiller. Two main zones of operation are identified: the high-pressure and the low-pressure zones. At low pressure, the refrigerant boils at a low temperature in the evaporator, taking heat from the supply water or air stream. The supply stream temperature is decreased (sometimes below 3°C), producing cooling to feed a cooling network. On the other hand, the refrigerant vapour goes through the compressor, rising its pressure and, consequently, its dew-point temperature. The refrigerant vapour is then cooled down and condensed in the condenser by a secondary cooling stream with a temperature close to ambient temperature. This secondary stream takes heat subtracted from the refrigerant to a cooling tower or radiator to be released into the atmosphere. The refrigerant, now in a liquid state, reduces its pressure when passing through an expansion valve and goes to the evaporator, completing the cycle.

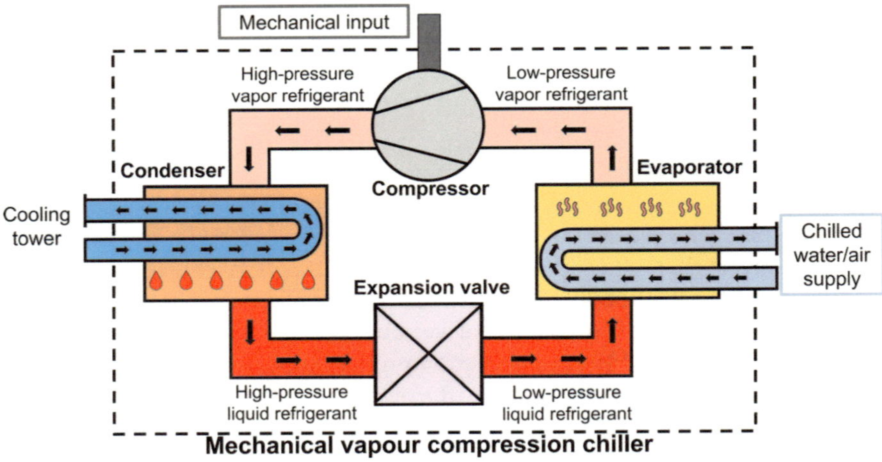

Figure 5.
Refrigeration cycle of a mechanical vapour compression chiller.

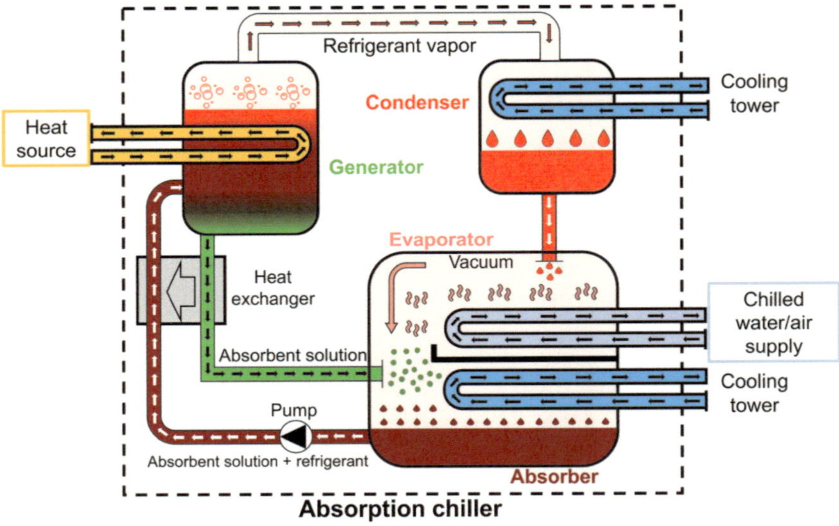

Figure 6.
Refrigeration cycle of a vapour absorption chiller.

3.2.2 Vapour absorption chillers

Unlike mechanical chillers, absorption chillers are driven by heat and do not employ a compressor or an expansion valve. Instead, they use two heat exchangers and an absorbent solution (e.g. lithium bromide) to modify the refrigerant's pressure.

Figure 6 illustrates the refrigeration cycle of an absorption chiller. This starts in the generator, which stores a mixture of a refrigerant and a concentrated absorbent solution. An external heat source heats the mixture up to the refrigerant's boiling temperature, separating the refrigerant vapour from the absorbent. The vapour goes to the condenser, where a secondary cooling stream extracts heat from the refrigerant to condense it. The condensed refrigerant and the absorbent solution from the generator enter the evaporator-absorber. Then, the refrigerant boils at a low temperature in the evaporator (due to close to vacuum conditions), producing cooling. This low-pressure condition results from the absorption process carried out in the absorber, where the absorbent strongly attracts the refrigerant vapour. Aided by a secondary stream, the absorbent-refrigerant mixture is condensed and pumped into the generator, completing the cycle. To increase efficiency, a heat exchanger is added between the streams of absorbent solution and absorbent-refrigerant mixture.

3.2.3 Vapour adsorption chillers

The refrigeration cycle of vapour adsorption chillers is characterised by the use of adsorbers, such as silica gel and activated carbon, which are solid substances that attract refrigerant vapour that sticks to their surface (i.e. adsorption).

Figure 7 shows a schematic of this kind of chillers. To achieve a continuous cycle, a minimum of two adsorbent beds are required, where one acts as an adsorber and the other as a desorber. The cycle starts with one of the beds already saturated with condensed refrigerant. The saturated bed is heated by an external heat source, desorbing the refrigerant by evaporation. A directional valve connects the bed with the condenser, which, supported by an external stream, condenses the refrigerant. Once in a liquid state, the refrigerant goes through an expansion valve, reducing its pressure until a nearly vacuum condition exists in the evaporator. Within the evaporator, the refrigerant boils at low temperature, producing cooling. The refrigerant vapour is adsorbed by the second bed, connected to the evaporator by another directional valve, and cooled down by a secondary stream. The cycle is completed by inverting the directional valves and heating the second adsorbent bed.

3.2.4 Coefficient of performance

Chiller efficiency is measured by a dimensionless CoP, which is the ratio of useful energy output for cooling to the primary energy input [33, 34]. For electric chillers,

$$CoP_{elect} = \frac{Q_{cool}}{W_{elec}} \tag{2}$$

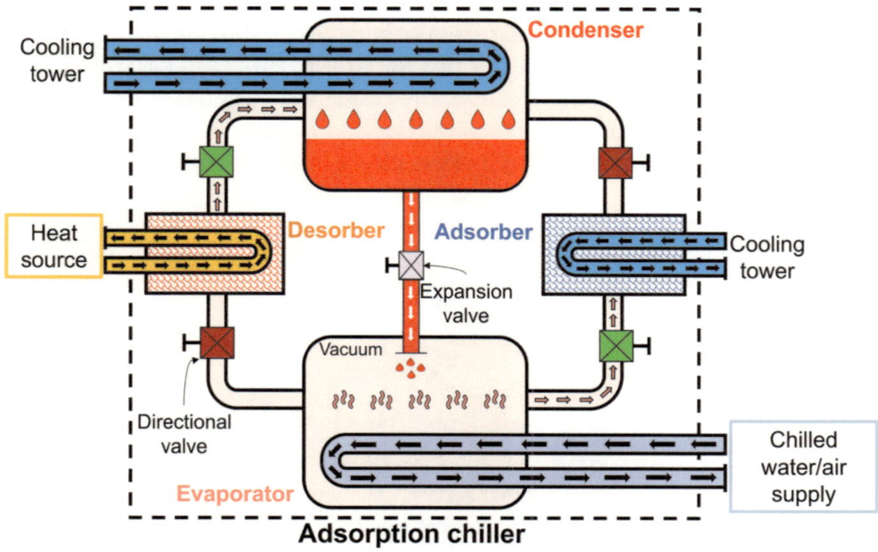

Figure 7.
Refrigeration cycle of a vapour adsorption chiller.

where Q_{cool} is the ratio of cooling rate capacity in the evaporator and W_{elec} is the electrical power input to the compressor. For absorption and adsorption chillers,

$$CoP_{abs/ads} = \frac{Q_{cool}}{Q_{in}} \tag{3}$$

where Q_{in} is the rate of heat input in the generator or desorber.

3.2.5 Refrigerant considerations

Before 1990, refrigerants were mainly produced from chlorofluorocarbons—ozone-depleting substances responsible for the "ozone hole". The Montreal Protocol, agreed in 1987, initiated a gradual phasing out of these substances for replacement with refrigerants with zero ozone-depleting potential [35]. In addition, the increasing concern on global warming led to the definition of the global warming potential (GWP) index, which is used to compare the effects of different greenhouse gases [36].

Refrigerants currently in use are classified as fluorocarbon and non-fluorocarbon refrigerants. Non-fluorocarbon refrigerants, such as ammonia, hydrocarbon, CO_2 and especially water have gained interest for their low GWP.

3.3 Co-generation and tri-generation power plants

3.3.1 Combined heat and power plants

Co-generation or CHP generation is the process of recovering the waste heat from electric power generation to produce steam or hot water. CHP plants are one of the main interfaces between electrical and thermal grids. Their configuration may vary, but waste heat recovery boilers are normally used. CHP plants can use an internal combustion engine or be based on a Brayton cycle (e.g. gas turbine), Rankine cycle, or a fuel cell [31, 37–39].

Figure 8 shows the configuration of a reciprocating engine CHP plant, which takes heat from the combustion gases and an engine block to generate hot water. Gas turbine CHP plants also take heat from exhaust gases, but steam is produced due to the high operating temperatures. In a Rankine cycle-based CHP plant, heat is recovered from the low-pressure steam in the condenser to generate hot water. The steam produced by a gas turbine CHP plant can be subsequently used in a Rankine cycle CHP plant to generate additional electricity. The integration of these thermodynamic cycles into a CHP plant is known as a combined cycle power plant [31, 37].

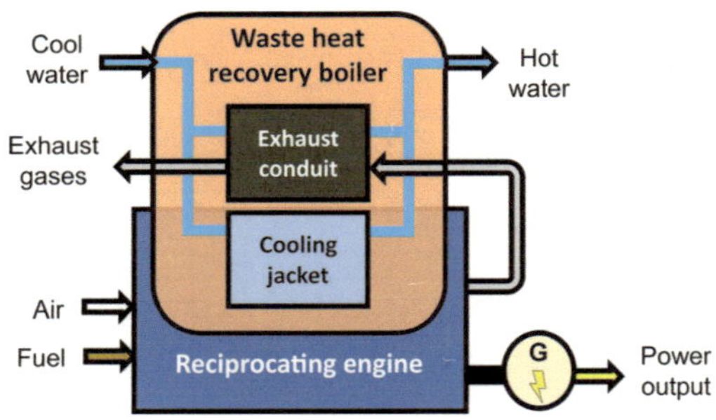

Figure 8.
Reciprocating engine co-generation plant.

Figure 9 illustrates the configuration of independent Brayton and Rankine cycle CHP plants.

In fuel cell-based CHP plants, instead of transforming thermal energy into mechanical energy to drive the shaft of a generator, fuel cells generate electrical power via an electrochemical reaction, which makes them more efficient. A schematic of a fuel cell stack working with hydrogen and oxygen is presented in **Figure 10**. Here, hydrogen (H_2) injected through the anode is separated into electrons and protons. Hydrogen protons travel through an electrolyte membrane to the cathode of an individual cell, where they combine with oxygen (O_2) and hydrogen electrons to produce water and heat. The electrical current produced by the flow of hydrogen electrons is forced through an electric circuit connecting anodes with cathodes of the fuel cells (i.e. current collector) to produce electrical power.

The operating temperature of a fuel cell may range from 80–1000°C depending on the electrolyte used [39]. Because of this, a cooling water system is required. Cooling water is circulated through the fuel cell stack, collecting heat and generating steam or hot water as a by-product of the process.

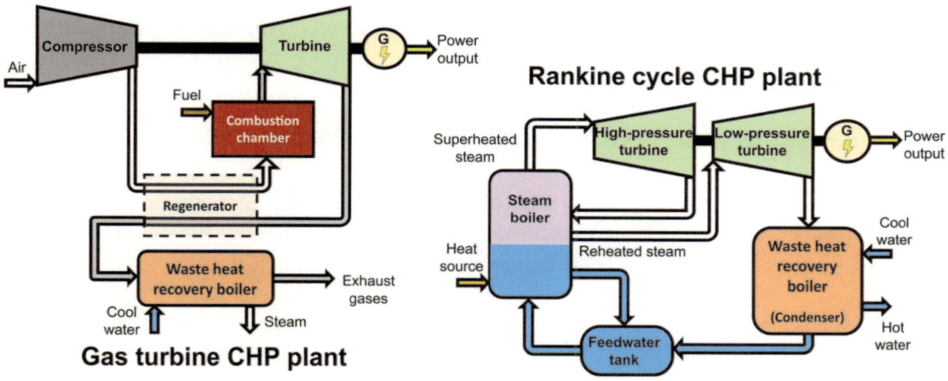

Figure 9.
Brayton and Rankine cycle co-generation plants.

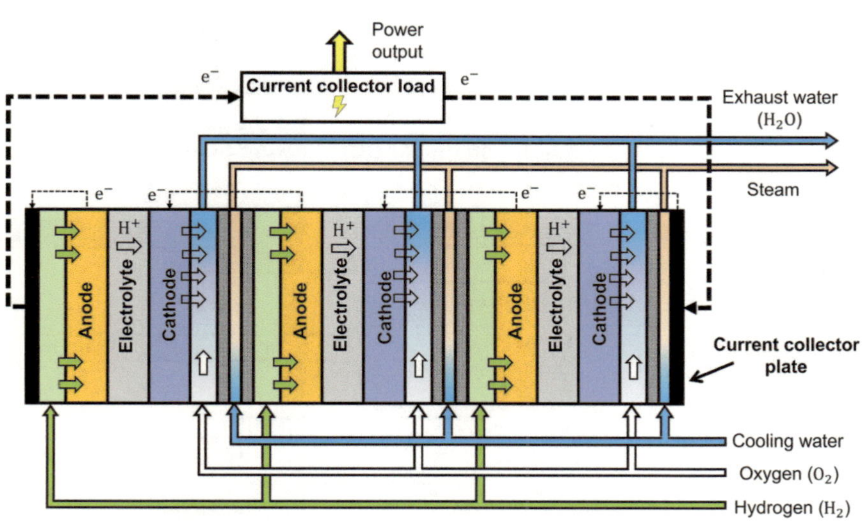

Figure 10.
Fuel cell-based co-generation plant.

3.3.2 Combined cooling, heat and power plants

A special case of co-generation is tri-generation, or combined cooling, heat and power (CCHP) generation. CCHP plants employ some electricity or heat generated by a CHP unit to produce chilled water. To this end, electrical or heat-driven chillers are used. Schematics illustrating the operation of CCHP plants are shown in **Figure 11**.

3.3.3 Control of co-generation power plants

Sensors and control loops are used in CHP plants to regulate their operation. Although the selection of control variables depends on the type of CHP plant, they may be either electrical power or heat-driven [40].

For electrical power led CHP plants, the control of electricity production is prioritised while heat is a by-product. The electrical power output is controlled and the water flow feeding the waste heat recovery boiler produces hot water at a specific temperature. Conversely, heat production is prioritised in heat led CHP plants, where the generated electricity is a by-product. A constant hot water flow at constant temperature is achieved in the waste heat recovery boiler by modifying the electrical power generation.

3.3.4 Performance indicators of co-generation and tri-generation power plants

In a co-generation power plant, the total input energy E_{in} is transformed into electrical energy W_{elec} and heating energy Q_{heat}. Performance of this process can be described in terms of electrical efficiency η_{elec} and thermal efficiency η_{th}:

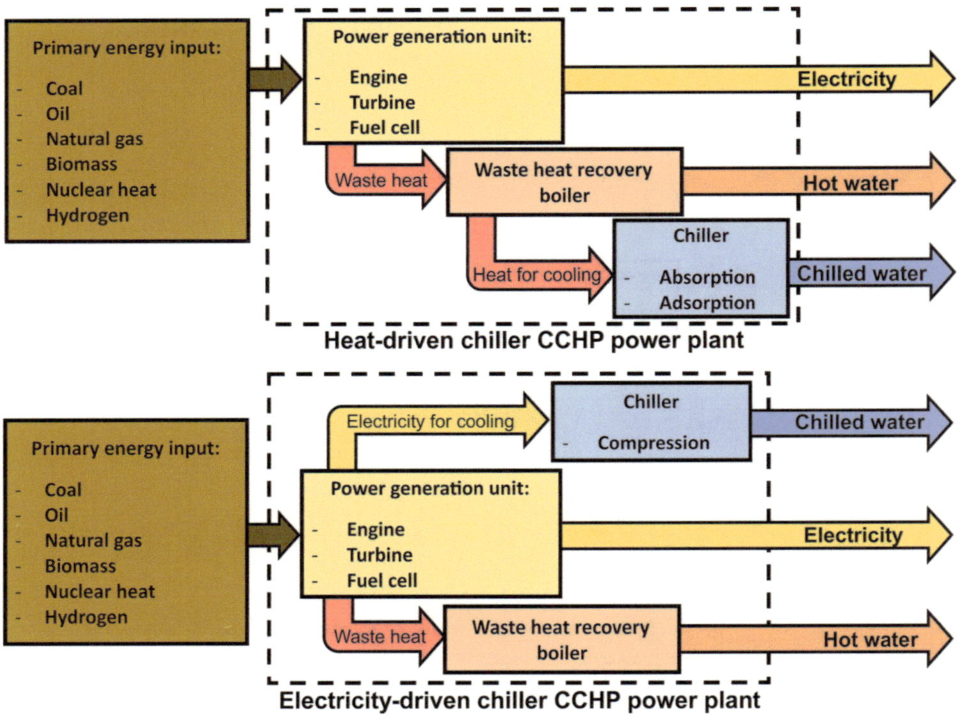

Figure 11.
Operation of tri-generation power plants.

$$\eta_{elect} = \frac{W_{elec}}{E_{in}}, \eta_{th} = \frac{Q_{heat}}{E_{in}} \tag{4}$$

The overall co-generation performance is normally quantified in primary energy savings (PES) achieved in comparison to the independent generation of electricity and heat [41, 42]. Although PES calculation may depend on national legislation [41], a direct calculation based on electrical and thermal efficiencies is given by

$$\text{PES}_{CHP} = \left(1 - \frac{1}{\eta_{elec,CHP}/\eta_{elec,SP} + \eta_{th,CHP}/\eta_{th,SP}}\right) \times 100 \tag{5}$$

where $\eta_{elec,CHP}$ and $\eta_{th,CHP}$ are the electrical and thermal efficiencies of the CHP plant, and $\eta_{elec,SP}$ and $\eta_{th,SP}$ the efficiencies for sperate production of electricity and heat.

Similarly, the performance of tri-generation plants is described by the cooling efficiencies of tri-generation $\eta_{cool,CCHP}$ and separate cooling production $\eta_{cool,SP}$:

$$\text{PES}_{CCHP} = \left(1 - \frac{1}{\eta_{elect,CCHP}/\eta_{elec,SP} + \eta_{th,CCHP}/\eta_{th,SP} + \eta_{cool,CCHP}/\eta_{cool,SP}}\right) \times 100 \tag{6}$$

where $\eta_{cool,CCHP}$ and $\eta_{cool,SP}$ depend on the type of chiller included in the CCHP plant (i.e. electrical, absorption or adsorption chiller). Effectively, the cooling efficiency of a CCHP plant is the product of the chiller's CoP and the electrical or thermal efficiency of the CHP unit. The calculations of $\eta_{cool,CCHP}$ and $\eta_{cool,SP}$ are given by

$$\eta_{cool,CCHP,elec} = CoP_{elec,CHP} \times \eta_{elec,CHP}, \ \eta_{cool,CCHP,abs/ads} = CoP_{abs/ads,CHP} \times \eta_{th,CHP} \tag{7}$$

$$\eta_{cool,SP,\ elec} = CoP_{elec,SP} \times \eta_{elec,SP}, \ \eta_{cool,SP,abs/ads} = CoP_{abs/ads,SP} \times \eta_{th,SP} \tag{8}$$

where subscripts $_{elec}$ and $_{abs/ads}$ denote the type of chiller.

3.4 Renewable energy sources and waste thermal energy recovery

3.4.1 Renewable electricity

Heating and cooling accounted for 49% of the total global energy consumption by the end of 2018, with an estimated 2% of this energy supplied by electricity from renewables [43]. RESs have gained importance in power production over the last decade, accounting for 27% of the global power generation in 2019 and 90% of the global power capacity increase in 2020. Although hydropower is still dominant, wind and solar photovoltaic (PV) are currently the leading technologies in growth.

Integrating electricity from RESs into thermal networks is aided by electric boilers and chillers. Another technology widely used is the electric heat pump—an essential component in new generations of DHC. Heat pumps work under the same operation principle as chillers and are either compression, absorption, or adsorption-based technologies. The main difference between a heat pump and a chiller is the ability of heat pumps to supply both cooling and heating. This is achieved by reversing the cycle in vapour compression chillers (see **Figure 5**) or exchanging the streams in the evaporator and condenser in absorption and adsorption chillers (**Figures 6** and 7). Thus, heat pumps may supply heating during winter and cooling during summer.

3.4.2 Geothermal heat and deep lake water cooling

The soil and local natural water reservoirs are important sources of thermal energy for heating and cooling. DHC systems can exploit existing local resources to significantly reduce electricity and gas consumption from national grids to drive boilers and chillers.

The heat produced at the Earth's core from nuclear reactions flows towards the surface, causing the ground temperature to vary with depth at an approximately constant rate of 30°C/km. This geothermal heat is exploited by extracting hot water from aquifers or using heat pumps to obtain heat from the shallow depth soil.

Although large power plants can use high temperature groundwater reservoirs to supply electricity to national grids, groundwater at temperatures below 100°C can be used for direct heating in local networks [44]. Heat exchangers are employed to subtract heat from the groundwater stream before it is reinjected to the ground.

For direct soil heat extraction or "geo-exchange", ground source heat pumps are used [45]. These consist of an electric heat pump connected to a water-filled pipe circuit buried into the ground to work as a heat exchanger between the soil and a water stream. Near the surface, the soil temperature varies significantly from season to season, but it remains relatively constant below 10 m [46]. Considering the ground temperature is higher than the outdoor temperature during winter and lower over the summer, the reversed operation capability of heat pumps is ideal to supply heating or cooling depending on the season of the year. These systems have proven to be suitable for heating cities and small communities in Denmark [47, 48].

Deep lake water cooling is the term adopted to identify the use of deep lakes, rivers or oceans to obtain free cooling from natural water bodies. Like groundwater systems, deep lake water cooling is achieved through a close or open heat transfer, using either heat exchangers or subtracting water directly from the bottom of the water bodies where the temperature is relatively constant. Examples include the district cooling systems in Paris and Ontario [9, 49].

3.4.3 Solar thermal systems

Unlike solar PV, solar thermal systems convert solar radiation directly into thermal energy. Solar collector panels (shown in **Figure 12**) are used to this end.

Based on the structure of the solar collector panel, solar collectors can be either flat-plate or evacuated tube collectors [50]. These panels are used to capture the solar radiation in a heat transfer medium, commonly water, which is displaced to a TES tank when the fluid reaches a temperature setpoint. The heat stored in the TES tank is later used for either heating or cooling purposes.

Solar thermal systems can be used together with solar PV cells to form solar hybrid photovoltaic-thermal (PV-T) collectors [50, 51]. Hybrid PV-T collectors are basically solar PV power plants which exploit the waste heat of solar PV generation. Thus, a solar hybrid PV-T collector is essentially a solar co-generation plant.

3.4.4 Waste thermal energy recovery and renewables integration through thermal energy storage

Thermal energy demand may be in a wide range of supplying temperatures. For example, power generation or industrial processes require heat from high-temperature fluids but generate significant waste heat. This differs from space heating or heat-driven refrigeration systems. The temperature difference between processes produces a thermal spectrum of energy use [52]. By adopting a cascaded heat supply system, an efficient use of the available thermal energy can be planned,

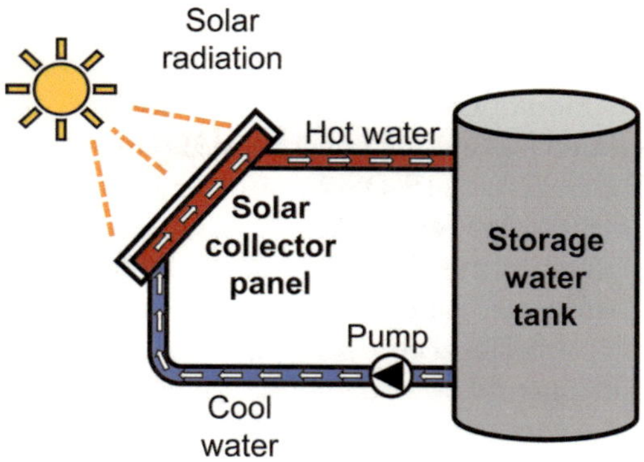

Figure 12.
Solar collector system.

where the waste thermal energy generated by a high-temperature range application can be recovered to supply other applications in a lower temperature range.

Industrial and commercial waste heat is an underutilised thermal energy resource with an enormous potential for DHC systems [53]. However, it is necessary to account for the temporary or geographical mismatch between waste thermal energy release and its demand. TES technologies are employed to this end, becoming crucial components when integrating recovered waste heat into thermal networks [54].

The use of TES elements is also important for the integration of RESs. Their intermittent nature is clearly exhibited in solar thermal systems, for which TES tanks are needed to collect as much energy as possible from the source while it is available. A further discussion on the different TES technologies can be found in Section 5.

4. Heating and cooling distribution technology

Distribution of heating and cooling from thermal energy sources to the final consumer can be achieved in a variety of manners. The distribution network and the ETS connecting a dwelling or installation are normally adapted to the needs of a customer and the availability of the sources. The different components and topologies of these two elements of DHC systems are revised next.

4.1 Distribution network

4.1.1 Pressure systems

The distribution network of a DHC system consists of an arrangement of underground pre-insulated pipelines filled with a heat transfer fluid (HTF). Although the vast majority use water as the HTF, CO_2-based DHC systems have been investigated [55, 56]. The distribution of thermal energy throughout the network involves a series of pressure systems to pump and effectively manage the thermal–hydraulic dynamics of the HTF.

The supply pressure system (SPS) is the primary system, which uses a pressure differential to move the HTF. It consists of several hydraulic pumps either centralised in a supply pressure station or distributed throughout the distribution network. A central supply pressure system (CSPS) is used for the HTF to provide a unidirectional flow path at each pipe segment of the network. Therefore, it is commonly adopted for networks with centralised thermal energy generation (CTEG) [57, 58]. Instead, a decentralised supply pressure system (DSPS) enables reversing the flow direction of the HTF. Two main DSPS configurations exist: a DSPS with distributed thermal energy generation (DTEG-DSPS) and a DSPS with distributed pumps at each ETS (ETS-DSPS) [10, 11, 59–63]. The flow paths for these systems are illustrated in **Figure 13**.

Most distribution networks are closed hydraulic circuits where the HTF follows a closed path in a confined space. Whenever the temperature of the HTF changes, the pressure of the whole network also changes. Expansion and pressurisation systems are used to avoid significant and potentially hazardous pressure changes [57, 58].

An expansion system contains several expansion vessels. These are filled with air or gas and HTF from the network, segregated by an expandable membrane to prevent them from mixing. The expansion system enables variations to the network's volume, where the distribution of thermal energy could be seen as a nearly isobaric process if the gas pressure does not significantly increase. The pressurisation system considers a mechanical pressurisation unit to guarantee a minimum pressure level in the network to prevent a phase change of the HTF at a low temperature.

4.1.2 Distribution network topologies

These refer to the pipe arrangement in a DHC system. Depending on the number of distribution lines, DHC systems can be one-pipe (1P), two-pipe (2P), three-pipe (3P) or four-pipe (4P) networks. In turn, these can be classified according to the configuration of the SPS and the directionality of the energy flow [64].

Conventional 3GDH and 3GDC systems are commonly 2P networks with a CSPS and a unidirectional thermal energy flow. Here, energy flow direction denotes the capacity of all ETSs to consume heating or cooling energy from the network. Systems with a bidirectional flow and a DSPS are instead associated with 5GDHC systems, where ETSs are active thermal energy producers or "prosumers" that supply and consume both heating and cooling. This results in a highly segmented network with different flow rates and temperatures at each pipe segment.

Open circuit networks without a return line are considered 1P networks. Here, the HTF is released directly to the environment instead of returning to the thermal energy source. Examples include networks directly supplied by rivers, lakes or geothermal water.

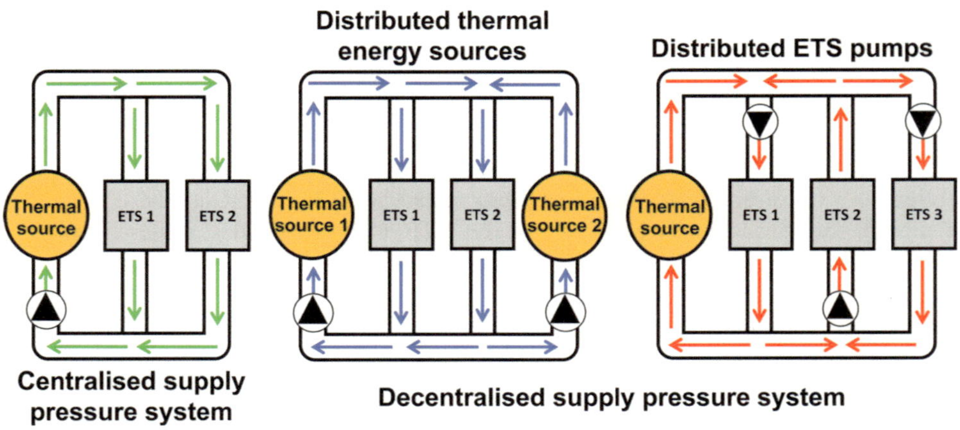

Figure 13.
Centralised and decentralised supply pressure systems.

Common return, recirculation and direct supply line connections are available for 3P networks, which aim to solve common problems with 2P networks. Common return line networks supply simultaneously heating and cooling through direct connections, incorporating a third line as a common return [65]. Recirculation of the supply line is necessary to guarantee a minimum supply temperature in 2P networks. To prevent significant thermal dissipation because of by-passing supply and return lines (due to their temperature difference), a direct recirculation line for the supply line is included [10, 11, 66, 67]. A direct supply line can be also added to 2P networks with operating temperatures close to ambient to supply heating or cooling with a higher or a lower temperature than in the other two pipelines [13, 64, 66].

4.2 Energy transfer stations

4.2.1 Station elements and connections

An ETS consists of two hydraulic circuits associated with the distribution network of the DHC system and the customers' installation. These circuits, primary and secondary, are linked via thermal–hydraulic couplers. The connections are either direct or indirect. An indirect connection is achieved by heat exchangers while a direct connection is obtained using hydraulic separators. Although direct connections are more efficient, indirect connections allow the primary and secondary sides to use a different HTF and prevent pollutants or the effects of damage or leakage to be transmitted from one side to the other [57].

Connections can be centralised or decentralised [22, 23]. In a centralised connection, a single heat transfer coupler is used. The thermal energy supplied from the primary to the secondary side is distributed through the customer's local installation. Decentralised connections use at least two heat transfer couplers: one for space heating or cooling and one for DHW.

ETSs may substantially differ in how thermal energy is subtracted from the distribution network. For ETS-DSPS networks, a local pump or a set of pumps creates a pressure differential to circulate HTF through the heat transfer couplers. Instead, for CSPS and DTEG-DSPS networks, a pressure boost is generated outside the ETS and control flow valves are placed at the inlet or outlet ports of the ETS. When a control flow valve opens, the pressure differential between the network's supply and return lines enables the HTF to flow. In bidirectional networks, this may require the use of a 3-way valve system to alternate between heating and cooling consumption [13]. Additionally, electric boilers, chillers and heat pumps may be used within the ETS to support the heating and cooling supply of a consumer.

A centralised ETS configuration for 3GDH systems is shown in **Figure 14** (left). This includes a central heat transfer coupling element for space heating and DHW production, a flow control valve in the outlet port, and a water tank for DHW on the secondary side. The diagram to the right shows an ETS-DSPS configuration typical of new generations of DHC systems. It includes a decentralised heat transfer connection for space heating or cooling and DHW, a heat pump and a 3-way valve system for flow reversal. Both ETSs include a heat meter to quantify thermal energy consumption.

4.2.2 Heat meters

Heat meters consist of temperature and flow sensors and an integration unit to calculate the thermal energy consumption of an ETS. The calculation is given by

$$E_{th} = \dot{V}\left(c_{p,sup}\rho_{sup}T_{sup} - c_{p,ret}\rho_{ret}T_{ret}\right)\Delta t \tag{9}$$

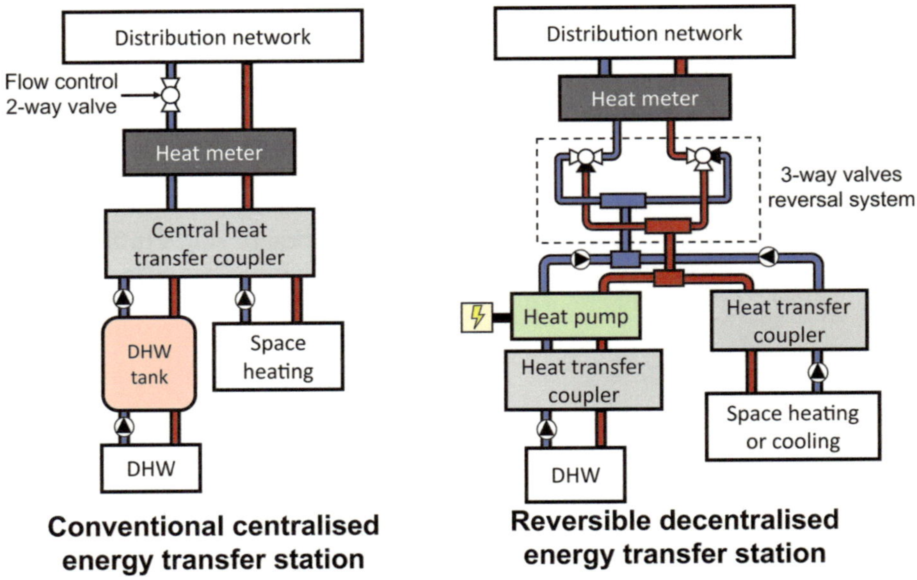

Figure 14.
Examples of energy transfer stations.

where the total thermal energy consumed E_{th} is obtained by integrating the heat flow difference between the supply and return lines at each time step Δt. In turn, such difference depends on the volumetric flow $\dot{V}$ passing through the ETS, the temperatures of the supply and return lines, T_{sup} and T_{ret}, and the specific heat value c_p and density ρ of the HTF corresponding to these temperatures [57, 58].

5. Thermal energy storage

An energy storage process can be considered in three stages: energy injection into the storage medium, energy holding and energy extraction, namely, charging, storing and discharging.

A wide variety of storage technologies have been developed. Their classification depends on the energy source (solar, geothermal, waste energy recovery), application (DHW, heating, cooling), location (centralised, distributed, underground, static, mobile), storage medium (air, water, soil, concrete, sand, salts), thermal energy collection system (active, passive), or the nature of thermal energy transfer couplers for charging or discharging storage devices (direct, indirect, both) [50, 54, 68–70].

5.1 Storage phenomena

5.1.1 Sensible heat storage systems

Sensible heat systems are based on the injection or extraction of heat into or from a storage medium to modify its enthalpy. Temperature levels are bounded to prevent a phase change of the medium. This enthalpy difference is called sensible heat.

The most common sensible heat TES units used in DHC systems are water-based [69]. They consist of insulated water tanks to prevent heat exchange between water and the environment during the storing stage and are used to store either heating or cooling energy.

The capacity of a medium to store thermal energy in the form of sensible heat is represented by its specific heat value c_p, which is the heat required to change the storage medium's temperature by 1°C per unit of mass. This relation is given by

$$\Delta H_{sens} = m_{st} c_p \Delta T_{st} \tag{10}$$

where a storage medium with a mass m_{st} exhibits a temperature difference ΔT_{st} resulting from a change in its enthalpy ΔH_{sens} [71].

5.1.2 Latent heat storage systems

Latent heat or phase change material (PCM) systems store more heat per unit of mass of the storage medium compared to sensible heat systems. Their application is widespread for cooling systems, where ice-liquid water storage is often used [69].

As in sensible heat systems, latent heat storage systems modify the enthalpy of the storage medium to store heat. As enthalpy changes by the addition or subtraction of heat, temperature of the medium also changes. This temperature variation stops at some point while the enthalpy keeps changing, resulting from a phase change of the medium. Such enthalpy difference at constant temperature is denominated latent heat, and it is reflected by a sudden change of the specific heat value of the medium.

Since latent heat storage systems consist of a two-phase storage medium, the enthalpy difference of the storage medium ΔH_{st} is equal to the addition of the sensible heat of both phases, namely ΔH_{sens}^{ph1} and ΔH_{sens}^{ph2}, and the latent heat H_{lat}:

$$\Delta H_{st} = \Delta H_{sens}^{ph1} + \Delta H_{sens}^{ph2} + H_{lat} \tag{11}$$

5.1.3 Thermochemical heat storage systems

These are classified depending on the phenomenon used for storage: chemical reactions and sorption processes. Through a reversible chemical reaction, a chemical is dissociated into and recombined from its components via endo and exothermic processes, storing and supplying heat accordingly. Absorption and adsorption bind a gas to an absorbent solution or a solid adsorber when heat is subtracted and release it when heat is added. These phenomena are represented by

$$R_1 + \text{Heat} \rightleftarrows R_2 + R_3 \tag{12}$$

where R_2 and R_3 are the gas and sorbent substance involved in sorption processes or the chemical products of a dissociative endothermic chemical reaction of R_1.

5.2 State-of-charge and storage efficiency

The total thermal energy H_{st} stored in a TES unit is estimated via temperature measurements at different points within the unit, discretising the storage volume in n nodes with an assumed uniformly distributed temperature [72, 73]:

$$H_{st} = \sum_{i=1}^{n} V_{st,i} \rho_i c_{p,i} T_{st,i} \tag{13}$$

This implies a summation of the internal energy of all nodes, where $V_{st,i}$ is the volume of the i-th node, $T_{st,i}$ the temperature measured in the node, and ρ_i and $c_{p,i}$ the density and the specific heat value of the storage medium at a temperature $T_{st,i}$.

Although H_{st} represents the total energy stored, only a portion of this energy can be used, limiting the storage capacity to minimum and maximum levels. The state-of-charge (SoC) is a dimensionless unit of the instantaneous available thermal energy in a TES unit, commonly given in percentage of maximum available energy or a value between 0 and 1 [74]. The SoC for a unit storing heating energy (SoC_{st}) is given by

$$SoC_{st} = \frac{H_{st} - H_{st}^{min}}{H_{st}^{cap}} = \frac{H_{st} - H_{st}^{min}}{H_{st}^{max} - H_{st}^{min}} \tag{14}$$

where H_{st} is the actual energy stored, H_{st}^{min} the minimum energy condition the unit can be discharged to, H_{st}^{max} the maximum energy condition the unit can be charged to, and H_{st}^{cap} the storage capacity of the unit. To calculate the SoC of a TES unit storing cooling energy, minimum and maximum energy terms are exchanged in (14).

Storing thermal energy entails an inherent energy loss. Thus, the performance of TES devices can be expressed in terms of the efficiency η_{st} of the energy storage process [69, 73]. For a generic TES unit storing heating energy, this is given by

$$\eta_{st} = \frac{H_{dis}}{H_{ch}} = \frac{H_{ch} - H_{loss}}{H_{ch}} \tag{15}$$

where H_{dis} is the energy subtracted during the discharging stage and H_{ch} the energy injected during the charging stage. The difference between the injected and subtracted energy is the energy loss H_{loss} during the storing stage.

6. Operation of district heating and cooling systems

6.1 Conventional control approach

DHC systems with CTEG and CSPS configurations are operated by a control system consisting of four cascaded control loops: the marginal pressure differential, supply temperature, flow, and thermal demand loops [57, 58, 67]. The marginal pressure differential control and the supply temperature control work in parallel at network level, whereas the flow control and the thermal demand control operate at each ETS. **Figure 15** illustrates the four control loops in a conventional district heating system.

The marginal pressure differential control ensures the CSPS supplies enough differential pressure to meet the flow rate demand of HTF of all ETSs connected to the grid. It maintains a constant pressure differential between supply and return lines at the farthest point of the distribution network. The supply temperature control regulates the thermal energy flow feeding the distribution network from the generation plant to maintain a specific supply temperature setpoint. Thus, these two control loops ensure sufficient thermal energy is supplied to all ETSs in the network by feeding them with a flow rate of HTF at a constant temperature.

The flow control regulates the opening of the flow control valve to modify the flow rate of the HTF passing through the primary side of each ETS. As the HTF is fed at a constant temperature, its flow rate is directly proportional to the flow of

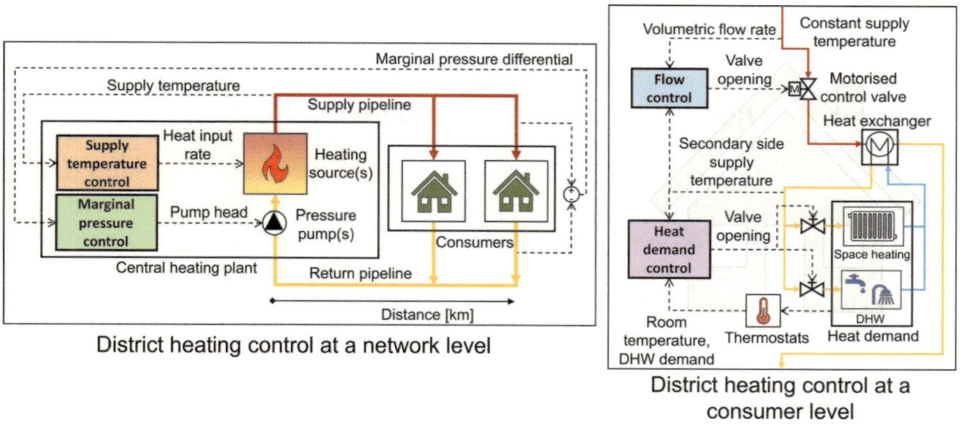

Figure 15.
Control of a conventional district heating system.

thermal energy entering the ETS. The control variable is selected depending on the type of thermal–hydraulic coupler used in each ETS. For ETSs working with heat exchangers, this is commonly the supply temperature at the secondary side [57, 58].

The thermal demand control is located at the lowest level of the cascaded control scheme, guaranteeing the thermal demand of each consumer is met according to the requirements for space heating, space cooling, DHW generation, or any other industrial or commercial process. As flow control guarantees a supply temperature of the ETS's secondary side, the requirements for thermal energy are regulated by the control of the final consuming elements within the consumer's premises (e.g. thermostatic radiator valves, flow control valves of DHW tanks, taps, showers).

6.2 Advanced control

Improvements to the conventional control approach previously discussed have been achieved. For instance, a supply temperature reset strategy to vary the supply temperature setpoint according to the ambient temperature increases the generation plant efficiency while decreasing thermal losses [57, 58]. Implementation of DSPS schemes, either DTEG-DSPS or ETS-DSPS, relieves the need for marginal pressure control [59–63]. In general, advanced control schemes have been a milestone for DHC systems as they mathematically optimise system operation through the coordinated management of thermal energy generation, storage, and distribution.

Advanced control schemes consider thermal load control as an additional element of a conventional control approach to optimise system operation [75]. Thermal load control is based on the operational flexibility of a thermal system and discussed next.

6.2.1 Operational flexibility

The stability of an energy network depends on the energy balance between supply and demand. This balance must be always fulfilled to prevent a deterioration of the system's performance. However, since the devices supplying and consuming thermal energy may cannot be operated as quickly or as frequently as the thermal energy demand changes, achieving energy balance is not easy. For example, the ramp-up or ramp-down of large thermal energy generation plants is slow, while the availability of RESs depends on weather conditions. Although some small-scale auxiliary devices can help maintain the energy balance, this comes at the expense of a higher utilisation of the different consuming or supplying devices in the network.

A definition of operational flexibility is provided in [75] as the ability to accelerate or delay the injection or extraction of energy into or from a system. For a thermal system, operational flexibility is mainly provided by energy storage elements. Although network pipelines and buildings have been used to provide flexibility [76, 77], the primary provision comes from TES units [78]. This concept is shown in **Figure 16**. TES units deal with the variations of thermal loads by storing thermal energy during periods when the supply is higher than demand and supplying this energy when the demand is higher than supply. The techniques to deal with load variations are known as load shaping [79].

6.2.2 Goals and challenges of advanced control

Advanced control focuses on the operational optimisation of a DHC system, which requires an optimal generation and distribution of thermal energy [75]. This entails:

- Selecting the most suitable generation method and prioritising energy dispatch from RESs against generation units fired by fossil fuels (thus supporting the decarbonisation of DHC systems).

- Implementing load shaping techniques to handle thermal load variations in the network and the intermittency of RESs.

- Decreasing operating temperatures in the distribution network to ambient temperature (or increasing for cooling) to reduce thermal losses.

- Provision of ancillary services by operating devices coupling electricity and thermal networks (i.e. CHP plants, electric boilers, compression chillers, heat pumps) to support balancing electrical energy generation and demand [80].

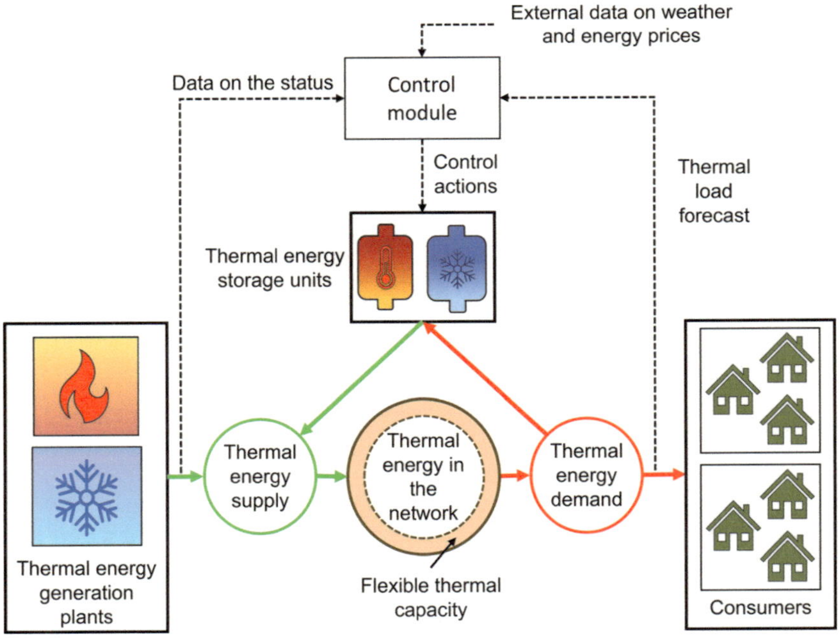

Figure 16.
Thermal load control based on operational flexibility.

The number of ETSs in DHC systems, the variability of thermal loads, thermal and electricity generation prices, and the complex dynamics in thermal networks constitute the main challenges preventing the adoption of advanced controls. DHC systems with several ETSs require many sensors to monitor and control the network's operation, increasing costs. Developing state estimators and new control strategies to reduce the amount of data may be key to relieve this challenge.

For an optimal operation of DHC systems, information on relevant operational variables is required. Temperatures and flow rates can be directly sensed and controlled. Still, the operation of thermal energy sources and TES units must be scheduled ahead for handling uncertainties on energy generation prices and thermal loads. Forecasts of prices and weather conditions, and models for determining the consumers' thermal loads are used to this end [81, 82].

The design of control systems requires a good understanding of the system's dynamics. Accurate and yet simple dynamic models of all the elements in the thermal networks are necessary, including generation plants, pipelines, storage units and thermal loads. These models can be used to reproduce the dynamic characteristics of the network (i.e. temperatures, pressures, flow rates, transport delays) and for quantifying the total operational flexibility available to operate the system [83–87].

6.2.3 Control configurations

Advanced control is categorised as central, distributed and hybrid control [75]. This classification depends on how the decision-making entities which operate the system are configured. In a central control topology, a centralised module interacts with all consumers, receiving online information of relevant operational variables. The central module combines this information with forecasts of system dynamics to generate and send control orders to optimise the operation of the system.

Distributed control, on the contrary, is based on multiple-decentralised modules operating the system. A multi-agent system architecture is commonly adopted [88, 89], where each agent may represent an element or a subsystem in the DHC network (e.g., producer, consumer, cluster of consumers, an energy storage plant). Agents communicate with each other, receiving information on the overall network conditions to optimise their operation locally.

Hybrid control integrates central and distributed control configurations. A central module optimises the operation of the entire system, tracing the conditions for an optimal thermal load in the network. This information is distributed to agents, which trade with each other to operate as close as possible to the optimal conditions [90].

6.2.4 Optimal operation

To guarantee an optimal operation of thermal networks, an optimisation process is performed in the short term by a central control module. This can be either an offline operational optimisation [91], or online model predictive control [92].

For DHC systems with distributed thermal energy sources, optimisation involves selecting energy sources to supply the network and quantify the best power output for each—known as unit commitment and economic dispatch, respectively [93, 94]. This should consider the flexibility provided by TES units to reduce costs [92].

Supply temperature control is a common practice to optimise system operation, aiming to reduce thermal losses in the network [91]. However, no specific control method has been firmly established for 5GDHC systems due to the highly variable working temperatures. A common approach bounding temperatures in the supply and return lines to maximum and minimum values has been adopted [12, 95].

7. Summary

In this chapter, DHC systems were described, including DHC system classification, technologies for thermal energy generation, distribution and energy storage, and topologies for distribution networks. In addition, the different components of thermal loads and the control strategies for the operation of a DHC system were reviewed.

The development and implementation of the latest generations of DHC systems is seen as an integral solution to satisfy local thermal energy demand. The potential of these systems to reduce carbon emissions by integrating local RESs and recovery of thermal energy from industrial and commercial activities, and an optimal energy generation and distribution represent a solid pathway to support the transition to sustainable local energy systems that many countries are aiming for.

The adoption of the thermal energy sources available locally, the increasingly complex dynamics resulting from implementing new topologies of ETSs and distribution networks, and the uncertainty in energy generation prices, local thermal loads variation, and intermittency of RESs make the optimal control of DHC systems highly challenging. Besides, the increasing number of coupling devices between the electricity grid and the local thermal networks requires a whole systems approach. A coordinated operation of local energy networks with different energy vectors and national energy systems is required, supporting each other in balancing energy supply and demand.

To deal with the challenges of implementing DHC systems, the development of mathematical models, analysis tools, forecasting methods, and the implementation of advanced controls are crucial to predict and study the dynamic interactions of the systems and to exploit the operational flexibility provided by TES units.

Abbreviations

5GDHC	5th generation of district heating and cooling
CCHP	Combined cooling, heat and power
CHP	Combined heat and power
CoP	Coefficient of performance
CSPS	Central supply pressure system
CTEG	Centralised thermal energy generation
DHC	District heating and cooling
DHW	Domestic hot water
DSPS	Decentralised supply pressure system
DTEG	Distributed thermal energy generation
ETS	Energy transfer station
GWP	Global warming potential
HTF	Heat transfer fluid
nGDC	n-th generation of district cooling
nGDH	n-th generation of district heating
PES	Primary energy savings
PV	Photovoltaic

PV-T	Photovoltaic-thermal
RES	Renewable energy source
SPS	Supply pressure system
TES	Thermal energy storage

Author details

Iván De la Cruz and Carlos E. Ugalde-Loo*
Cardiff University, Wales, UK

*Address all correspondence to: ugalde-looc@cardiff.ac.uk

References

[1] Werner S. District heating and cooling. In: Scott A. Elias et al., editors. Reference Module in Earth Systems and Environmental Sciences. Elsevier; 2013. DOI: 10.1016/B978-0-12-409548-9.01094-0

[2] Wilson C, Jones N, Devine-Wright H, Devine-Wright P, Gupta R, Rae C, Tingey M. EnergyRev. Common types of local energy system projects in the UK [Internet]. 2020. Available from: https://www.energyrev.org.uk/media/1442/energyrev_archetypesreport_202009.pdf

[3] Thellufsen J, Lund H. Roles of local and national energy systems in the integration of renewable energy. Applied Energy. 2016;183;419-429. DOI: 10.1016/j.apenergy.2016.09.005

[4] Østergaard P. Reviewing optimisation criteria for energy systems analyses of renewable energy integration. Energy. 2009;34(9); 1236-1245. DOI: 10.1016/j.energy.2009.05.004

[5] Collins J. The history of district heating. District Heating. 1959; 44(4):154-61.

[6] Lund H, Werner S, Wiltshire R, Svendsen S, Thorsen J, Hvelplund F, Mathiesen B. 4th Generation District Heating (4GDH): Integrating smart thermal grids into future sustainable energy systems. Energy. 2014;68:1-11. DOI: 10.1016/j.energy.2014.02.089

[7] Steiner A, Yumkella K, Clos J, Begin G. District Energy in Cities: Unlocking the Potential of Energy Efficiency and Renewable Energy. United Nations Environment Programme (UNEP): Nairobi, Kenya [Internet]. 2015. Available from: https://wedocs.unep.org/20.500.11822/9317

[8] Paris climate action plan towards a carbon neutral city, 100% renewable energies, resilient, fair and inclusive [Internet]. 2018. Available from: https://cdn.paris.fr/paris/2020/11/23/257b26474ba3ba08ee02baa096f9c5dd.pdf

[9] CLIMESPACE. Sustainable cooling in the heart of the city [Internet]. 2018. Available from: https://www.climespace.fr/storage/publications/pdf/sustainable-cooling-in-the-heart-of-the-city.pdf

[10] Averfalk H, Werner, S. Novel low-temperature heat distribution technology. Energy. 2018; 145:526-539. DOI: 10.1016/j.energy.2017.12.157

[11] Lund H, Østergaard P, Chang M, Werner S, Svendsen S, Sorknæs P, Möller B. The status of 4th generation district heating: Research and results. Energy. 2018;164:147-159. DOI: 10.1016/j.energy.2018.08.206

[12] Pass R, Wetter M, Piette M. A thermodynamic analysis of a novel bidirectional district heating and cooling network. Energy. 2018;144:20-30. DOI: 10.1016/j.energy.2017.11.122

[13] Buffa S, Cozzini M, D'Antoni M, Baratieri M, Fedrizzi R. 5th generation district heating and cooling systems: A review of existing cases in Europe. Renewable and Sustainable Energy Reviews. 2019;104:504-522. DOI: 10.1016/j.rser.2018.12.059

[14] Boesten S, Ivens W, Dekker S, Eijdems H. 5th generation district heating and cooling systems as a solution for renewable urban thermal energy supply. Advances in Geosciences. 2019;49:129-136. DOI: 10.5194/adgeo-49-129-2019

[15] Gadd H., Werner S. Heat load patterns in district heating substations. Applied energy. 2013;108:176-183. DOI: 10.1016/j.apenergy.2013.02.062

[16] Gadd H, Werner S. Daily heat load variations in Swedish district heating

systems. Applied Energy. 2013;106: 47-55. DOI: 10.1016/j.apenergy.2013.01.030

[17] Noussan M, Jarre M, Poggio A. Real operation data analysis on district heating load patterns. Energy. 2017;129: 70-78. DOI: 10.1016/j.energy.2017.04.079

[18] Chu C, Jong T, Huang Y. Thermal comfort control on multi-room fan coil unit system using LEE-based fuzzy logic. Energy Conversion and Management. 2005;46(9-10):1579-1593. DOI: 10.1016/j.enconman.2004.08.002

[19] Källblad K. Thermal Models of Buildings. Determination of temperatures, heating and cooling loads. Theories, models and computer programs [thesis]. Lund University; 1998.

[20] Werner S. The heat load in district heating systems [thesis]. Chalmers tekniska högskola; 1984.

[21] Lin Y, Stout J, Yu V, Vidic R. Disinfection of water distribution systems for Legionella. In: Seminars in respiratory infections; 1998; Saunders. 13(2); p. 147-159.

[22] Elmegaard B, Ommen T, Markussen M, Iversen J. Integration of space heating and hot water supply in low temperature district heating. Energy and Buildings. 2016;124:255-264. DOI: 10.1016/j.enbuild.2015.09.003

[23] Yang X, Li H, Svendsen S. Alternative solutions for inhibiting Legionella in domestic hot water systems based on low-temperature district heating. Building services engineering research and technology. 2016;37(4):468-478. DOI: 10.1177/0143624415613945

[24] Difs K, Danestig M, Trygg L. Increased use of district heating in industrial processes–impacts on heat load duration. Applied Energy. 2009;86(11):2327-2334. DOI: 10.1016/j.apenergy.2009.03.011

[25] Ilic D, Trygg L. Economic and environmental benefits of converting industrial processes to district heating. Energy conversion and management. 2014;87:305-317. DOI: 10.1016/j.enconman.2014.07.025

[26] Vaghefi A, Farzan F, Jafari M. Modeling industrial loads in non-residential buildings. Applied Energy. 2015;158:378-389. DOI: 10.1016/j.apenergy.2015.08.077

[27] Brückner S, Liu S, Miró L, Radspieler M, Cabeza L, Lävemann E. Industrial waste heat recovery technologies: An economic analysis of heat transformation technologies. Applied Energy. 2015;151:157-167. DOI: 10.1016/j.apenergy.2015.01.147

[28] Danielewicz J, Śniechowska B, Sayegh M, Fidorów N, Jouhara H. Three-dimensional numerical model of heat losses from district heating network pre-insulated pipes buried in the ground. Energy. 2016;108: 172-184. DOI: 10.1016/j.energy.2015.07.012

[29] Basu P, Kefa C, Jestin L. Boilers and burners: design and theory. Springer Science & Business Media; 2012. DOI: 10.1007/978-1-4612-1250-8

[30] Mikkanen P, Jokiniemi J, Kauppinen E, Vakkilainen E. Coarse ash particle characteristics in a pulp and paper industry chemical recovery boiler. Fuel. 2001;80(7):987-999. DOI: 10.1016/S0016-2361(00)00195-2

[31] Sipilä K. Cogeneration, biomass, waste to energy and industrial waste heat for district heating. In: Wiltshire R, editor. Advanced District Heating and Cooling (DHC) Systems. Woodhead Publishing; 2016. p. 45-73. DOI: 10.1016/B978-1-78242-374-4.00003-3

[32] Heselton P. Boiler operator's handbook. 2nd ed. River Publishers; 2020. 488 p. DOI: 10.1201/9781003151876

[33] Gordon J, Ng K, Chua H. Centrifugal chillers: thermodynamic modelling and a diagnostic case study. International Journal of refrigeration. 1995;18(4):253-257. DOI: 10.1016/0140-7007(95)96863-2

[34] Kuczyńska A, Szaflik W. Absorption and adsorption chillers applied to air conditioning systems. Archives of Thermodynamics. 2010;31(2):77-94. DOI: 10.2478/v10173-010-0010-0

[35] Olama A. District cooling: Theory and practice. CRC Press; 2016. 304 p. DOI: 10.4324/9781315371634

[36] Shine K. The global warming potential—the need for an interdisciplinary retrial. Climate Change. 2009;96:467-472. DOI: 10.1007/s10584-009-9647-6

[37] Dinçer İ, Bicer, Y. Integrated Energy Systems for Multigeneration. Elsevier; 2019. 453 p. DOI: 10.1016/C2015-0-06233-5

[38] Safa H. Heat recovery from nuclear power plants. International Journal of Electrical Power & Energy Systems. 2012;42(1):553-559. DOI: 10.1016/j.ijepes.2012.04.052

[39] Okada O, Yokoyama K. Development of polymer electrolyte fuel cell cogeneration systems for residential applications. Fuel Cells. 2001;1(1):72-77. DOI: 10.1002/1615-6854(200105)1:1<72::AID-FUCE72>3.0.CO;2-P

[40] Diangelakis N, Pistikopoulos E. Modelling, design and control optimization of a residential scale CHP system. In: Kopanos G, Liu P, Georgiadis M, editors. Advances in Energy Systems Engineering. Springer, Cham; 2017. p. 475-506. DOI: 10.1007/978-3-319-42803-1_16

[41] Cardona E, Piacentino A. Cogeneration: a regulatory framework toward growth. Energy Policy. 2005;33(16):2100-2111. DOI: 10.1016/j.enpol.2004.04.007

[42] Chicco G, Mancarella P. Trigeneration primary energy saving evaluation for energy planning and policy development. Energy policy. 2007;35(12):6132-6144. DOI: 10.1016/j.enpol.2007.07.016

[43] IEA. Renewable Energy Policies in a Time of Transition: Heating and Cooling [Internet]. 2020. Available from: https://iea.blob.core.windows.net/assets/66a547ea-09bf-48fb-b7bb-6b94ef80b083/Renewable_Energy_Policies_in_a_Time_of_Transition_-_Heating_and_Cooling.pdf

[44] Bloomquist R. Geothermal space heating. Geothermics. 2003;32(4-6);513-526. DOI: 10.1016/j.geothermics.2003.06.001

[45] Omer A. Ground-source heat pumps systems and applications. Renewable and sustainable energy reviews. 2008;12(2):344-371. DOI: 10.1016/j.rser.2006.10.003

[46] Hanova J, Dowlatabadi H. Strategic GHG reduction through the use of ground source heat pump technology. Environmental Research Letters. 2007;2(4):044001. DOI: 10.1088/1748-9326/2/4/044001

[47] Østergaard P, Lund H. A renewable energy system in Frederikshavn using low-temperature geothermal energy for district heating. Applied Energy. 2011;88(2);479-487. DOI: 10.1016/j.apenergy.2010.03.018

[48] Østergaard P, Mathiesen B, Möller B, Lund H. A renewable energy scenario for Aalborg Municipality based on low-temperature geothermal heat,

wind power and biomass. Energy. 2010; 35(12);4892-4901. DOI: 10.1016/j.energy.2010.08.041

[49] Boyce F, Hamblin P, Harvey L, Schertzer W, McCrimmon R. Response of the thermal structure of Lake Ontario to deep cooling water withdrawals and to global warming. Journal of Great Lakes Research. 1993;19(3):603-616. DOI: 10.1016/S0380-1330(93)71244-7

[50] Sarbu I, Sebarchievici C. Solar heating and cooling systems: Fundamentals, experiments and applications. Academic Press; 2016. 424 p. DOI: 10.1016/B978-0-12-811662-3.00003-7

[51] Herrando M, Pantaleo A, Wang K, Markides C. Solar combined cooling, heating and power systems based on hybrid PVT, PV or solar-thermal collectors for building applications. Renewable Energy. 2019;143:637-647. DOI: 10.1016/j.renene.2019.05.004

[52] Fox D, Sutter D, Tester J. The thermal spectrum of low-temperature energy use in the United States. Energy & Environmental Science. 2011;4(10): 3731-3740. DOI: 10.1039/C1EE01722E

[53] Fang H, Xia J, Zhu K, Su Y, Jiang Y. Industrial waste heat utilization for low temperature district heating. Energy policy. 2013;62:236-246. DOI: 10.1016/j.enpol.2013.06.104

[54] Miró L, Gasia J, Cabeza L. Thermal energy storage (TES) for industrial waste heat (IWH) recovery: A review. Applied energy. 2016;179:284-301. DOI: 10.1016/j.apenergy.2016.06.147

[55] Weber C, Favrat D. Conventional and advanced CO2 based district energy systems. Energy. 2010;35(12): 5070-5081. DOI: 10.1016/j.energy.2010.08.008

[56] Henchoz S, Weber C, Maréchal F, Favrat D. Performance and profitability perspectives of a CO2 based district energy network in Geneva's City Centre. Energy. 2015;85:221-235. DOI: 10.1016/j.energy.2015.03.079

[57] Skagestad B, Mildenstein P. District heating and cooling connection handbook [Internet]. NOVEM, Netherlands Agency for Energy and the Environment, 2002. Available from: https://www.districtheatingscotland.com/wp-content/uploads/2015/12/DistrictHeatingAndCoolingConnectionHandbook.pdf

[58] Euroheat & Power. Guidelines for District Heating Substations [Internet]. 2008. Availablre from: https://www.euroheat.org/wp-content/uploads/2008/04/Euroheat-Power-Guidelines-District-Heating-Substations-2008.pdf

[59] Gu J, Wang J, Qi C, Yu X, Sundén B. Analysis of a hybrid control scheme in the district heating system with distributed variable speed pumps. Sustainable Cities and Society. 2019;48: 101591. DOI: 10.1016/j.scs.2019.101591

[60] Yan A, Zhao J, An Q, Zhao Y, Li H, Huang Y. Hydraulic performance of a new district heating systems with distributed variable speed pumps. Applied energy. 2013;112:876-885. DOI: 10.1016/j.apenergy.2013.06.031

[61] Kuosa M, Kontu K, Mäkilä T, Lampinen M, Lahdelma R. Static study of traditional and ring networks and the use of mass flow control in district heating applications. Applied Thermal Engineering. 2013;54(2):450-459. DOI: 10.1016/j.applthermaleng.2013.02.018

[62] Laajalehto T, Kuosa M, Mäkilä T, Lampinen M, Lahdelma R. Energy efficiency improvements utilising mass flow control and a ring topology in a district heating network. Applied thermal engineering. 2014;69(1-2): 86-95. DOI: 10.1016/j.applthermaleng.2014.04.041

[63] Wang H, Wang H, Zhu T. A new hydraulic regulation method on district

heating system with distributed variable-speed pumps. Energy Conversion and Management. 2017;147: 174-189. DOI: 10.1016/j.enconman.2017.03.059

[64] Sulzer M, Hangartner D. Grundlagen-/Thesen Kalte Fernwärme (Anergienetze) [Internet]. 2014. Available from: https://www.researchgate.net/publication/262418798_Grundlagen-Thesen_Kalte_Fernwarme_Anergienetze

[65] Frederiksen S, Werner S. District Heating and Cooling. Lund, Sweden: Studentlitteratur AB, Lund; 2013. 586 p.

[66] Dalla Rosa A, Li H, Svendsen S. Method for optimal design of pipes for low-energy district heating, with focus on heat losses. Energy. 2011;36(5): 2407-2418. DOI: 10.1016/j.energy.2011.01.024

[67] Averfalk H, Werner S. Essential improvements in future district heating systems. Energy Procedia. 2017;116: 217-225. DOI: 10.1016/j.egypro.2017.05.069

[68] Lizana J, Chacartegui R, Barrios-Padura A, Valverde J. Advances in thermal energy storage materials and their applications towards zero energy buildings: A critical review. Applied Energy. 2017;203:219-239. DOI: 10.1016/j.apenergy.2017.06.008

[69] Guelpa E, Verda V. Thermal energy storage in district heating and cooling systems: A review. Applied Energy. 2019;252:113474. DOI: 10.1016/j.apenergy.2019.113474

[70] Alva G, Lin Y, Fang G. An overview of thermal energy storage systems, Energy. 2018;144:341-378. DOI: 10.1016/j.energy.2017.12.037

[71] Kalaiselvam S, Parameshwaran R. Thermal energy storage technologies for sustainability: systems design, assessment and applications. Elsevier; 2014. 430 p. DOI: 10.1016/C2013-0-09744-7

[72] Barz T, Seliger D, Marx K, Sommer A, Walter S, Bock H, Körkel S. State and state of charge estimation for a latent heat storage. Control Engineering Practice. 2018;72:151-166. DOI: 10.1016/j.conengprac.2017.11.006

[73] Han Y, Wang R, Dai Y. Thermal stratification within the water tank. Renewable and Sustainable Energy Reviews. 2009;13(5):1014-1026. DOI: 10.1016/j.rser.2008.03.001

[74] Abutayeh M, Alazzam A, El-Khasawneh B. Optimizing thermal energy storage operation. Solar Energy. 2015;120:318-329. DOI: 10.1016/j.solener.2015.06.027

[75] Vandermeulen A, van der Heijde B, Helsen L. Controlling district heating and cooling networks to unlock flexibility: A review. Energy. 2018;151: 103-115. DOI: 10.1016/j.energy.2018.03.034

[76] Kensby J, Trüschel A, Dalenbäck J. Potential of residential buildings as thermal energy storage in district heating systems–Results from a pilot test. Applied Energy. 2015;137: 773-781. DOI: 10.1016/j.apenergy.2014.07.026

[77] Merkert L, Haime A, Hohmann S. Optimal scheduling of combined heat and power generation units using the thermal inertia of the connected district heating grid as energy storage. Energies. 2019;12(2):266. DOI: 10.3390/en12020266

[78] Nuytten T, Claessens B, Paredis K, Van Bael J, Six D. Flexibility of a combined heat and power system with thermal energy storage for district heating. Applied energy. 2013;104: 583-591. DOI: 10.1016/j.apenergy.2012.11.029

[79] Arteconi A, Hewitt N, Polonara F. State of the art of thermal storage for demand-side management. Applied Energy. 2012;93:371-389. DOI: 10.1016/j.apenergy.2011.12.045

[80] Salpakari J, Mikkola J, Lund P. Improved flexibility with large-scale variable renewable power in cities through optimal demand-side management and power-to-heat conversion. Energy Conversion and Management. 2016;126:649-661. DOI: 10.1016/j.enconman.2016.08.041

[81] Heller A. Heat-load modelling for large systems. Applied Energy. 2002;72 (1):371-387. DOI: 10.1016/S0306-2619 (02)00020-X

[82] Fang T, Lahdelma R. Evaluation of a multiple linear regression model and SARIMA model in forecasting heat demand for district heating system. Applied energy. 2016;179:544-552. DOI: 10.1016/j.apenergy.2016.06.133

[83] Schweiger G, Larsson P, Magnusson F, Lauenburg P, Velut S. District heating and cooling systems–Framework for Modelica-based simulation and dynamic optimization. Energy. 2017;137:566-578. DOI: 10.1016/j.energy.2017.05.115

[84] Lauster M, Teichmann J, Fuchs M, Streblow R, Mueller D. Low order thermal network models for dynamic simulations of buildings on city district scale. Building and Environment. 2014; 73:223-231. DOI: 10.1016/j.buildenv.2013.12.016

[85] Jie P, Tian Z, Yuan S, Zhu N. Modeling the dynamic characteristics of a district heating network. Energy. 2012; 39(1):126-134. DOI: 10.1016/j.energy.2012.01.055

[86] Leśko M, Bujalski W. Modeling of district heating networks for the purpose of operational optimization with thermal energy storage. Archives of thermodynamics. 2017;38(4): 139-163. DOI: 10.1515/aoter-2017-0029

[87] Bøhm B, Seung-kyu Ha, Won-tae Kim, Koljonen T, Larsen HV, Lucht M et al. Simple models for operational optimisation. Sittard: Netherlands Agency for Energy and the Environment, 2002. 135 p. (IEA District Heating and Cooling, Annex VI: Report 2002:S1).

[88] Wernstedt F. Multi-Agent Systems for Distributed Control of District Heating Systems [Internet] [PhD dissertation]. [Karlskrona]: Blekinge Institute of Technology; 2005. (Blekinge Institute of Technology Doctoral Dissertation Series). Available from: http://urn.kb.se/resolve?urn=urn:nbn:se:bth-00317

[89] van Deventer J, Gustafsson J, Delsing J. Controlling district heating load through prices. In: 2011 IEEE International Systems Conference; 4-7 April 2011; Montreal. Canada: IEEE; 2011. p. 461-465. DOI: 10.1109/SYSCON.2011.5929104

[90] Bünning F, Wetter M, Fuchs M, Müller D. Bidirectional low-temperature district energy systems with agent-based control: Performance comparison and operation optimization. Applied Energy. 2018;209:502-515. DOI: 10.1016/j.apenergy.2017.10.072

[91] Benonysson A, Bøhm B, Ravn H. Operational optimization in a district heating system. Energy conversion and management. 1995;36(5):297-314. DOI: 10.1016/0196-8904(95)98895-T

[92] Verrilli F, Srinivasan S, Gambino G, Canelli M, Himanka M, Del Vecchio C, Glielmo L. Model predictive control-based optimal operations of district heating system with thermal energy storage and flexible loads. IEEE Transactions on Automation Science and Engineering. 2016;14(2):547-557. DOI: 10.1109/TASE.2016.2618948

[93] Sandou G, Font S, Tebbani S, Hiret A, Mondon C, Tebbani S, Mondon C. Predictive control of a complex district heating network. Proceedings of the 44th IEEE Conference on Decision and Control. 2005:7372-7377. DOI: 10.1109/CDC.2005.1583351

[94] Wu C, Gu W, Jiang P, Li Z, Cai H, Li B. Combined economic dispatch considering the time-delay of district heating network and multi-regional indoor temperature control. IEEE Transactions on Sustainable Energy. 2017;9(1):118-127. DOI: 10.1109/TSTE.2017.2718031

[95] Wirtz M, Neumaier L, Remmen P, Müller D. Temperature control in 5th generation district heating and cooling networks: An MILP-based operation optimization. Applied Energy. 2021;288: 116608. DOI: 10.1016/j.apenergy.2021.116608

Chapter 5

Electricity Storage in Local Energy Systems

William Seward, Weiqi Hua and Meysam Qadrdan

Abstract

Traditionally, power system operation has relied on supply side flexibility from large fossil-based generation plants to managed swings in supply and/or demand. An increase in variable renewable generation has increased curtailment of renewable electricity and variations in electricity prices. Consumers can take advantage of volatile electricity prices and reduce their bills using electricity storage. With reduced fossil-based power generation, traditional methods for balancing supply and demand must change. Electricity storage offers an alternative to fossil-based flexibility, with an increase expected to support high levels of renewable generation. Electrochemical storage is a promising technology for local energy systems. In particular, lithium-ion batteries due to their high energy density and high efficiency. However, despite their 89% decrease in capital cost over the last 10 years, lithium-ion batteries are still relatively expensive. Local energy systems with battery storage can use their battery for different purposes such as maximising their self-consumption, minimising their operating cost through energy arbitrage which is storing energy when the electricity price is low and releasing the energy when the price increases, and increasing their revenue by providing flexibility services to the utility grid. Power rating and energy capacity are vitally important in the design of an electricity storage system. A case study is given for the purpose of providing a repeatable methodology for optimally sizing of a battery storage system for a local energy system. The methodology can be adapted to include any local energy system generation or demand profile.

Keywords: Battery sizing, flexibility, lithium-ion battery storage, local energy systems, trading mechanisms

1. Introduction

In the context of the power system operation, 'flexibility' can be defined as the capability of the power system to match demand and supply in the face of rapid and large swings in supply and/or demand. Traditionally, flexibility has been provided by large scale dispatchable power generating units such as gas and coal power generating plants. This type of flexibility is referred to as 'supply-side flexibility'. The increasing share of variable renewable sources of energy in power systems intensifies the challenge of balancing electricity supply and demand. However, the decommissioning of the fossil-based dispatchable generating units necessitates the use of alternative sources of flexibility to compensate for the variations of renewable power generation. Battery storage technologies are a key alternative source of

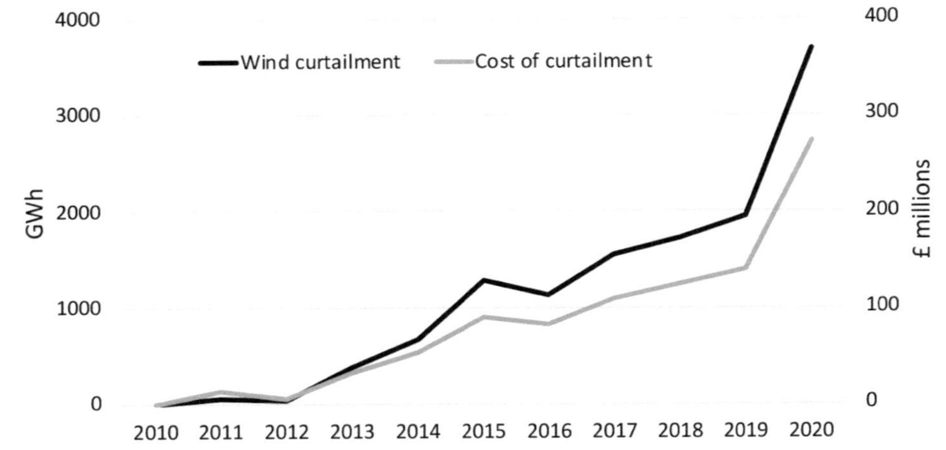

Figure 1.
Annual wind curtailment and cost of curtailment in Great Britain. This figure is produced using data from [1].

flexibility along with demand-side response and interconnectors with neighbouring power systems.

The increase of wind and solar generation capacity in the GB power system and the lack of sufficient flexibility in recent years have led to the curtailment of renewable energy. **Figure 1** shows the annual wind curtailment and the cost of curtailing wind energy in GB from 2010 to 2020. When the electricity generated by wind is high during the low demand period, and the system cannot absorb all the electricity generated by wind due to lack of storage, demand turn-up and downward generation capacity, a fraction of wind generation is curtailed to ensure the supply and demand are balanced and the system frequency is kept as close as possible to 50 Hz.

The variability of power outputs from renewable generation also has an impact on short term electricity prices in day ahead and intraday wholesale markets. At times when there is excess renewable electricity in the system, the electricity price decreases (sometimes it becomes negative) to encourage consumers to use more electricity and encourage generators to reduce their outputs. On the other hand, during high demand periods when the electricity generation by renewable is low, the electricity price rises to encourage consumers to consume less electricity. The increased variation in half-hourly electricity prices is a market signal for flexibility requirement and can make investment in electricity storage financially viable. Being able to shift electricity consumption, consumers that are able to shift their electricity consumption will be able to reduce their electricity bill.

For achieving the net-zero target in the UK by 2050, which was passed as a law by the British government in 2019 [2], National Grid suggested three scenarios namely '*System Transformation*'[1], '*Leading the Way*'[2] and '*Consumer Transformation*'[3]. These scenarios envisaged different mixes of technologies and measures to achieve the net zero target. According to National Grid's scenarios [3], the share of electrical energy generated from wind and solar is expected to increase from 33% in 2019 to between 74% (for System Transformation scenario) and 87% (for Leading

[1] ***System Transformation scenario*** emphasis on hydrogen for heating, consumers less inclined to change behaviour, lower energy efficiency, and supply side flexibility.

[2] ***Leading the Way scenario*** emphasis on fastest credible decarbonisation, significant lifestyle change, mixture of hydrogen and electrification for heating.

[3] ***Consumer Transformation scenario*** emphasis on electrified heating, consumers' willingness to change behaviour, high energy efficiency, and demand side flexibility.

the Way scenario) [3]. **Figure 2** shows that the significant increase in the capacity of wind and solar generation is expected to coincide with the reduction of the capacity of dispatchable generation. To ensure the future power system can balance supply and demand under various operating conditions, there will be an increasing role for electricity storage. According to the National Grid scenarios [3], by 2050, the capacity of electricity storage will increase from 3.75 GW in 2019 to 37.3 GW in Customer Transformation scenario, 23.5 GW in System Transformation scenario and 40.4 GW in Leading the Way scenario.

Figure 3 demonstrates how battery storage can be used to support the balancing of electricity supply and demand, and mitigate the variation of renewable

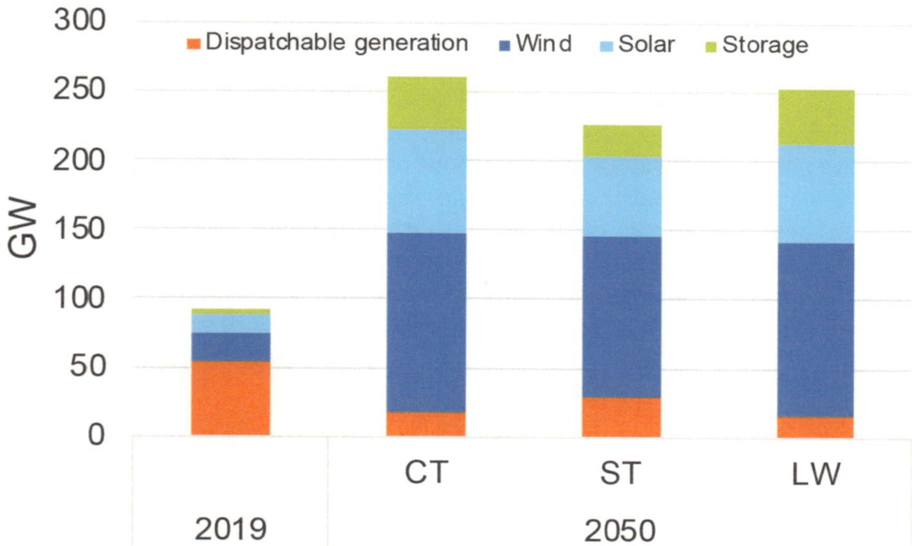

Figure 2.
Capacity of dispatchable generation plants, renewable and storage. Dispatchable generation in this figure accounts for nuclear, biomass, hydrogen and fossil power plants. Columns above year 2050 represent a range of future scenarios; CT: Consumer Transformation, ST: System Transformation, LW: Leading the Way. This figure was produced using data from [3].

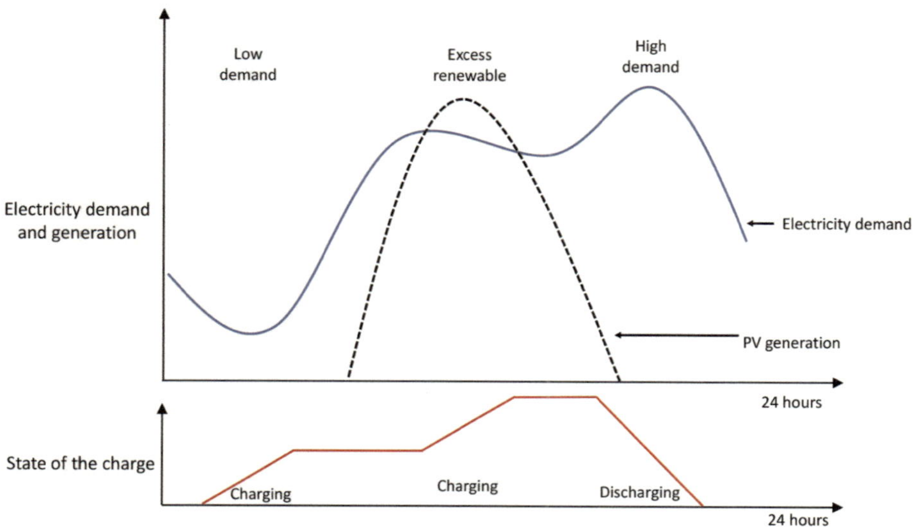

Figure 3.
A schematic depicting the application of battery storage to balance electricity demand and supply.

generation. One of the key uses of battery storage is to smooth the net electricity demand (i.e. total electricity demand minus renewable electricity generation). The battery storage is charged when electricity demand is low and also during periods when electricity from renewable is high. The stored electricity will be discharged during peak demand hours to reduce the stress on the power system.

2. Electricity storage technologies

There is a variety of electricity storage technologies that use a range of mediums to store energy. Electricity storage systems can store energy in mechanical, chemical, electrical and magnetic mediums. Several technologies have characteristics that make them suitable for use in local energy systems. This section identifies electricity storage technologies that are available and those that have characteristics that make them suitable for use in local energy systems. Detailed characteristics of each technology are given, as well as short descriptions of how they work.

There is a variety of existing technologies that can store electrical energy for later use. These have a range of size and operating characteristics that are summarised in **Table 1**. In the context of electricity storage for local energy systems, some of the technologies in **Table 1** are more suitable than the others. The technologies that are promising for applications in local energy systems are listed in the latter half of **Table 1**.

Technology	Power capacity	Energy capacity	Efficiency (%)	Response time	Lifetime (years, cycles)
Pumped hydro storage	100–3000 MW	Up to 100 GWh	75–85	Seconds to minutes	40–60 years
Compressed air energy storage (CAES)	5–1000 MW	100 MWh – 10 GWh	70–89	1–15 min	20–40 years, 13,000+
Flywheels	0.1–20 MW	10–100 kWh	70–95	ms - secs	>15 years, 100,000+
Capacitor	0–50 kW	0–50 kWh	60–65	ms	5, 50,000+
Supercapacitor	0–300 kW	0–300 kWh	90–95	ms	>20 years, 100,000+
Cryogenic [8]	100 kW – 300 MW	100 kWh – 2.4 GWh	40–60	Minutes	25+, 13,000+
Promising (battery) technologies for local energy systems					
Lead-acid	Some kW – 10 MW	Up to 10 MWh	75–90	5–10 ms	3–15, 500–3000
Sodium sulphur [9, 10]	50 kW – 34 MW	400 kWh – 58 MWh	80–90	1 ms	10–15, 2500–4500
Lithium ion	1 kW – 100 MW	Up to 10 MWh	85–98	10–20 ms	5–15, 1000–10,000
Nickel-cadmium [11, 12]	0–40 MW	Up to 13 MWh	60–65	ms	10–20, 2000–3500
Flow batteries (Vanadium and Zinc) [13]	30 kW – 10 MW	3.75 kWh – 32 MWh	75–85	< 1 ms	5–10, 2000–2012,000+

Table 1.
Characteristics of electricity storage technologies [4–7]. The technologies that are most suitable for local energy systems are identified.

Pumped hydro storage: pumped storage is a large-scale mechanical storage technology that stores electricity as potential energy. A difference in height is required between two large volumes of water. Water is pumped to the higher reservoir during charge and flows to the lower reservoir during discharge. Traditional pumped hydro storage is less suited to local energy systems due to its large-scale, high cost, long construction times and niche geographical requirements [4, 7, 14].

Compressed air energy storage: CAES stores electricity as highly pressurised air, in the form of potential energy. During the charging process, air is pumped into large storage caverns. At discharge, the pressurised air is released through a turbine to generate electricity. Similar to pumped hydro storage, the application of traditional CAES in local energy systems is limited by its large-scale, high cost and specific geographical requirements [4, 7, 15].

Flywheel storage: Flywheels store kinetic energy in a rotating mass or rotor. The rotor is situated within a low pressure (vacuum) chamber and connected to a motor/generator. During charging, the rotor speed increases, during discharge the rotor speed decreases. Flywheels have high power density with rapid response times. However, they have relatively low energy storage capacity and high self-discharge. These factors, as well as relatively high cost per kWh mean flywheels are less suited to provide flexibility in local energy systems. For further detail on flywheels, their design, operation and applications, refer to [16].

Capacitors: Capacitors consist of two electrical conductors with a thin insulating layer between them. When a capacitor is charged, it stores energy in an electrostatic field [14]. They have high power density but have limited energy capacity, low efficiency and relatively high self-discharge losses [7, 14, 17]. These factors make capacitors less suited to applications within local energy systems.

Supercapacitor: Supercapacitors are electrochemical double-layer capacitors [7, 14, 18]. In comparison to a traditional capacitor, supercapacitors provide higher energy densities but with lower power density [19]. Despite having a higher energy density than capacitors and better efficiency than most other storage technologies, supercapacitors still have relatively low energy density in comparison to other available technologies and higher levels of energy capacity are very costly. For further details on the state of the art of supercapacitors, refer to [20].

Cryogenic storage: During the charging process of cryogenic or 'liquid air' electricity storage, a cryogen (liquid air) is produced and stored in vacuum insulated tanks [21]. During the discharge process the cryogen is heated and boils. The boiled cryogen is sent through a cryogen heat engine which generates electricity. Cryogenic electricity storage has relatively high energy density, low capital cost per kWh and the potential for increased energy capacity with relative ease [7, 21]. However, cryogenic electricity storage has an efficiency range of less than 50%, making it significantly less efficient than alternative technologies [5, 22].

In the remainder of this section, the electrochemical (battery) storage technologies that show most promise for local energy system are discussed.

2.1 Electrochemical energy storage

Electrochemical energy storage can be split into two major categories, integrated energy storage systems and external energy storage systems [7]. Integrated energy storage systems have their electrochemical charging and discharging reactions taking place within the battery, with no spatial separation [7]. In contrast, flow batteries house their liquid electrolytes in separate containers, bringing them together for a reversible chemical reaction, which enables charging and discharging [7].

Integrated energy storage systems are common with battery technologies such as Lead-acid, Sodium-sulphur, Nickel-cadmium, Nickel-metal hydride and Lithium-ion [7, 23]. One of the major advantages of integrated energy storage systems is their scalability. A very wide range of power and energy capacities are possible as individual cells can be amalgamated into a larger system.

There is a variety of chemistries for integrated energy storage systems (common examples are given in **Table 1**). These chemistries differ in their chemical construction, resulting in differing storage characteristics.

Among the available integrated energy storage technologies, lithium batteries are of increasing importance for energy storage in recent years [24]. Their high specific energy and energy density make lithium batteries suitable for transport applications [25]. In addition to these characteristics, lithium batteries have long life cycles, low maintenance, scalability, power and energy rating flexibility, very high efficiency and low self-discharge [4, 7, 25].

Flow batteries are electrochemical energy storage systems that consist of two electrolytes separated by an ion-selective membrane, in an electrochemical cell. The electrolytes are stored in separate tanks, giving the advantage of total decoupling of power and energy ratings [4, 7, 17, 24]. The chemical reaction that occurs is entirely reversible [7]. The most common flow battery technologies are polysulphide bromide (PSB), vanadium redox (VRB) and zinc bromide (ZnBr) [24]. The flexibility in power and capacity rating, in addition to their scalability from adding additional storage tanks or electrochemical cells, makes flow batteries suitable for a range of applications. Flow batteries have fast response times and long service life, no degradation with deep discharge and low self-discharge [4, 7, 26]. However, while currently being in early commercialisation stage, drawbacks such as relatively low energy density and complex system requirements (including sensors, pumps, flow and power management) mean flow battery storage systems make up a very small proportion of current storage [6, 26–28].

2.2 Lithium battery storage systems

Lithium-ion (Li-ion) battery technology was first patented in 1982 and commercialised in 1991 [29]. Since then, lithium-ion has dominated battery technologies and has replaced nickel-cadmium (NiCd) batteries in mobile phones, doubling the energy density of on-board battery storage [29]. Multiple chemistries are commercially available (these are listed in **Table 2**) however they work in a similar way. Lithium ions move through the electrolyte between the anode and cathode. As the battery discharges, lithium ions are released from the anode and are diffused into the cathode. Anodes are typically made from graphite-based material due to the low cost and availability of carbon. Cathode materials are the dominant factor in determining energy storage performance and hence differentiates technologies. The battery chemistries in **Table 2** are listed by the material used for the cathode. For further details on anodes, cathodes and electrolytes and more details about how lithium-ion batteries work, refer to [32].

2.2.1 Advantages

Lithium-ion battery storage is the second most mature battery energy storage technology in the market, after lead acid batteries [30]. For a detailed review of lithium-ion chemistries, refer to [33].

Lithium-ion battery technology has a variety of advantages that make it a popular choice for portable electronic devices, where energy density, size, weight,

Chemical composition	Acronym	Common application
Lithium manganese oxide	LMO	Electric vehicles, consumer electronics
Lithium iron phosphate	LFP	Electric vehicles, consumer electronics, power tools, aviation
Lithium nickel manganese cobalt oxide	NMC	Electric vehicles, power tools, grid energy storage
Lithium nickel cobalt aluminium	NCA	Electric vehicles
Lithium cobalt oxide	LCO	Consumer electronics
Lithium titanate	LTO	Automotive and grid storage

Table 2.
Existing lithium-ion battery chemistries [25, 30, 31].

Advantages	Disadvantages
High energy density [4, 7, 17, 29, 34, 35] Longevity, long cycle life [6, 17, 29, 35] Versatility and scalability [4, 29] High efficiency [7, 14, 17, 34, 35] Rapid response [6, 14] Low self-discharge [7, 34, 35] Low operation and maintenance requirements [35, 36] Satisfactory operating temperature ranges [35] High reliability [35] Relatively fast recharge [35]	Highly flammable, fire hazard, safety hazard [25, 29, 30, 35] Highly sensitive to temperature [30, 36] Accelerated degradation during tough operating conditions [7, 14] Advanced battery management system required [7, 14, 35, 36] High initial cost [35] Potential for material bottleneck with high demand [35] Currently weak recovery and recycling schemes [35]

Table 3.
Advantages and disadvantages of lithium-ion battery storage.

longevity and efficiency are important. A summary of lithium-ion battery advantages and disadvantages are shown in **Table 3**.

NMC chemistries are the most typically used in grid-scale storage applications due to their balanced performance of energy, power, cost and cycle life [30]. One of the most attractive characteristics of lithium-ion batteries for stationary energy storage applications is their scalability. The ability to connect individual cells together to create an energy storage system with higher energy and power rating makes lithium-ion batteries very versatile. Creating storage systems in this manner means lithium-ion batteries can provide an extremely wide range of power and energy ratings, as shown in **Table 1**: power ratings range between 1 kW – 100 MW and energy ranges up to 10 MWh [4]. Another characteristic that makes lithium-ion battery storage well suited to local energy system electricity storage is the rapid response time. This allows local energy systems to respond quickly to changes in electricity price and market signals to alter their demand, which can allow them to participate in a wider range of markets.

2.2.2 Disadvantages

Lithium is highly flammable and is therefore a safety and fire hazard. Lithium-ion batteries are sensitive to temperature, requiring an active cooling system to maintain an optimal operating temperature. To ensure the safety of lithium-ion batteries, active battery management systems are required to track key parameters such as voltage, current, state of charge, state of health and temperature [35].

A common drawback of battery storage is degradation over time. As shown in **Table 1**, lithium-ion batteries have a long life. However, they are still subject to degradation that can be accelerated by their operating conditions. Batteries are subject to calendar and cycle aging [35, 37]. Calendar aging occurs when batteries are stored without cycling. Cycle aging is degradation related to charge and discharge. Both types of aging cause capacity fade and loss of power, leading to reduced energy and power capacity. Operating conditions such as temperatures outside of the recommended operating range, overcharging, deep discharging and high currents, typically accelerate battery degradation [35, 38]. Optimal operating temperatures depend on the manufacturer, typically these vary around 21°C [30]. Extended periods with the temperature too far from this can lead to loss of performance of the battery.

Typical battery degradation characteristics are non-linear. **Figure 4** shows a representation of the fall in energy capacity seen in batteries, which can be split into three sections. (1) being a rapid initial fall, (2) is a steady reduction and (3) is another rapid drop in energy capacity [37].

Batteries that are at 'end-of-life' are no longer suitable for their application and must be replaced. For high energy density applications such as electric vehicles and portable electronic devices, the battery end of life is typically 80% of its original energy capacity [35, 37]. For stationary storage applications, such as in local energy systems, energy density is not as vital. Therefore, batteries can be utilised to a lower level of degradation before replacement.

2.2.3 Sustainability and recyclability of lithium-ion batteries

Increasing use of lithium-ion batteries in transport and power applications will lead to high demand for battery minerals and large numbers of spent batteries. The three main options for correct disposal of lithium-ion batteries that have reached their end of life are remanufacturing, repurposing and recycling [39].

Remanufacturing is the process of refurbishing lithium-ion batteries and utilising them for the same application. This process is the most effective in terms of maximising the value of end-of-life lithium-ion batteries, as well as minimising life-cycle energy consumption and emissions. However, remanufacturing has relatively strict requirements such as an acceptable state of health and must meet regulatory requirements for power, energy, life cycle and others. For example, electric vehicle batteries that have less than 80% of their original energy capacity left are not suitable for their application.

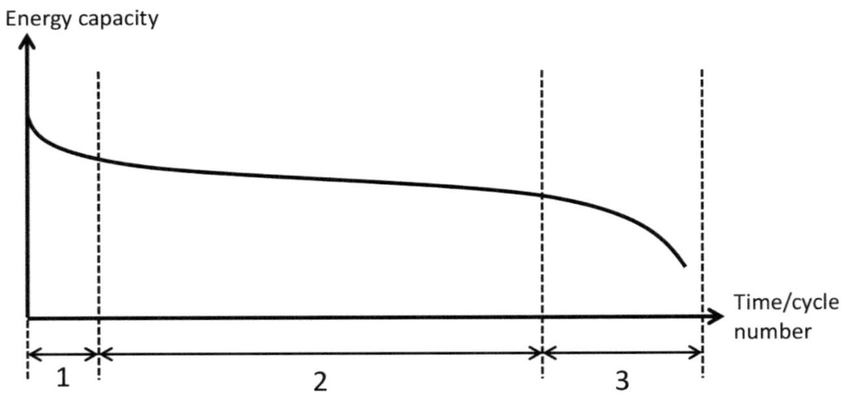

Figure 4.
Typical energy capacity degradation characteristic of a lithium-ion battery storage [37].

Repurposing is the re-deployment of end-of-life batteries for 'second life' applications where power and energy ratings are less critical. Where remanufacturing of spent lithium-ion batteries is not viable, repurposing is an attractive prospect, before recycling. Repurposing is typically the use of spent 'first life' electric vehicle batteries in stationary grid storage applications. However, repurposing of spent lithium-ion batteries faces challenges such as:

- Necessary replacement of damaged cells.
- Integration of a new battery management system that is compatible with the new application.
- Reliably grading of each battery pack.
- Being flexible enough to accommodate for a wide range of manufacturer designs, scales, compatibility and chemistries.

In addition to these challenges, repurposed battery storage systems must be price competitive with first-life batteries, despite complex processing requirements.

Recycling is the final stage of battery recyclability, which is vital for extracting critical materials from spent batteries, creating a circular economy and reducing demand for mining raw materials. Two technologies are commercially used to recycle end-of-life lithium-ion batteries. The first, pyrometallurgy, employs high temperatures to extract and purify metals inside the lithium-ion battery [40]. The second, hydrometallurgy, leaches the internal content of the battery to transfer metals from solid phase to the aqueous solution [40]. A combination of techniques can also be used. Direct recycling has also been demonstrated, using supercritical carbon dioxide for solvent extraction [41]. For further details of recycling methods and a comparison of methods, refer to [40, 41].

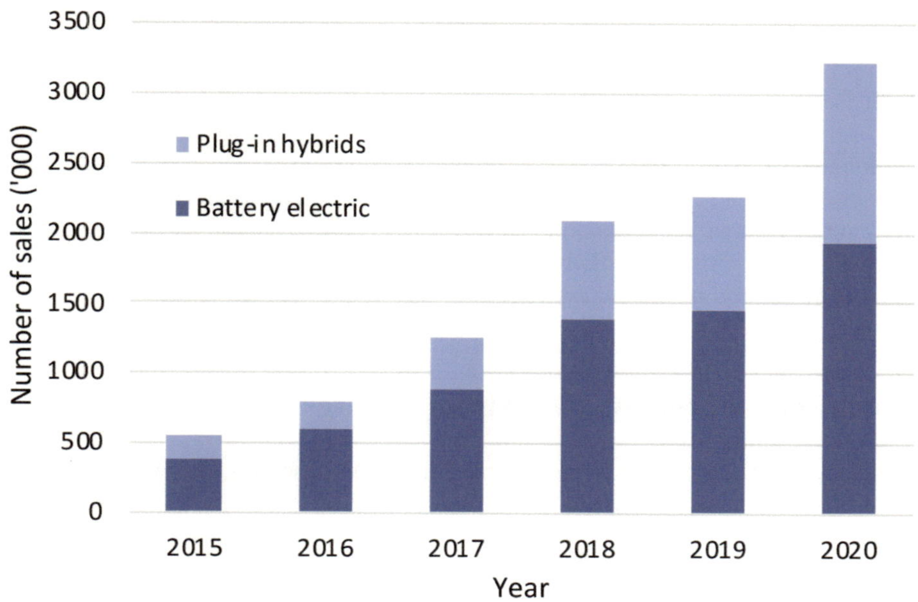

Figure 5.
Global plug-in electric vehicle sales from 2015 to 2020. Reproduced from [42].

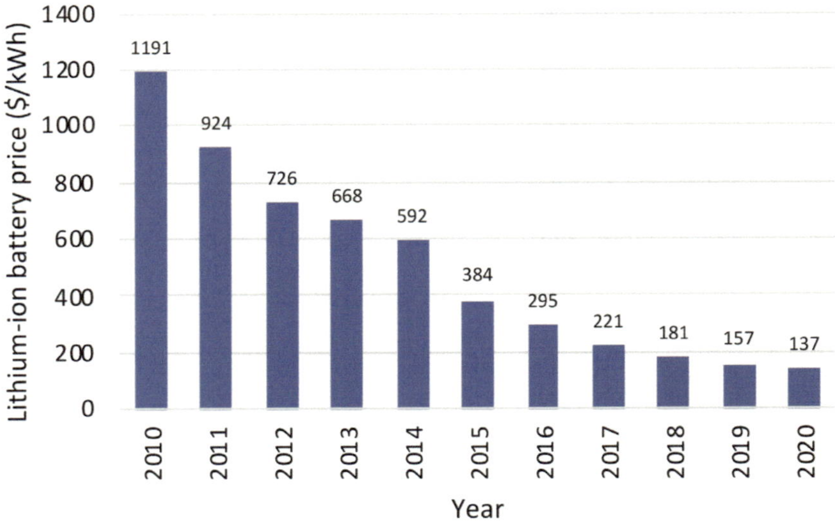

Figure 6.
Lithium-ion battery price, volume weighted average, all sectors from 2010 to 2020. Reproduced from [44].

2.2.4 Market for lithium-ion batteries

Lithium-ion batteries are the dominant storage technology for applications such as portable electronic devices and electric transport. In recent years, the increase in demand for battery electric vehicles has led an increase in lithium-ion battery demand. **Figure 5** shows the increase in purchased electric vehicles over the last 10 years.

The increase in demand for lithium-ion batteries for electric vehicles has contributed to an 89% reduction in cost over 10 years [43]. The BloombergNEF 2020 lithium-ion battery price survey reports on the price of lithium-ion batteries, which is shown in **Figure 6**.

Prices for lithium-ion batteries have been reported at under $100/kWh, with the average price reported as $137/kWh [43]. Although these low prices for lithium-ion batteries have been driven by demand in electric vehicles, there is significant benefit for stationary energy storage applications.

Lithium-ion batteries that are for stationary storage applications have the strongest global growth and market share, in comparison to other electrochemical storage technologies [45]. With global behind the meter battery storage expected to significantly grow in years to come [46]. Lower cost lithium-ion batteries enable local energy systems to affordably increase their flexibility, giving them the ability to control when and how much energy they exchange with the grid.

3. Applications of energy storage in local energy systems

Historically, the electricity system in GB has been based on a unidirectional flow of electricity from large scale power stations to consumers. Power is transported through a high voltage power transmission network, to medium and low voltage power distribution networks, where it is delivered to consumers. Emission reduction targets have driven the transition from this conventional centralised energy system to a decentralised and localised paradigm which integrates diverse distributed energy sources, e.g. solar photovoltaic (PV) and wind generation. This transformation encourages increasing numbers of local energy systems [47]. These local energy systems enable electricity consumers to become prosumers which can generate energy for self-consumption, energy trading with each other, and providing energy and flexibility services to the utility grid. However, the intermittency of renewable energy sources and mismatch between the on-site generation and consumption present a challenge for prosumers. Incorporating energy storage into the local energy systems provides a key solution for prosumers to flexibly manage the distributed energy sources and participate in local energy markets. Instead of curtailing the excessive generation from solar and wind, the surplus generation can be charged into the energy storage devices, and discharged during the peak demand periods.

3.1 Energy arbitrage

Energy arbitrage is affected by a consumer strategically storing the energy when the retail electrical energy price is low, and releasing the stored energy when the retail electrical energy price increases. Energy arbitrage is the most direct method for profiting from energy storage. An example of energy charge of a consumer under peak and off-peak tariffs is presented in **Table 4**. If this consumer stores additional 8 kWh electricity at the off-peak tariff (10 pence/kWh), the total charge would reduce from 258 pence to 130 pence.

Researchers in [48] demonstrated that by providing community battery energy storage systems to prosumers with solar PV and electric vehicles, prosumers could earn additional profits through the energy arbitrage. In [49], a home energy management system was developed for individual prosumers incorporating energy storage systems and solar PV in supporting demand response and energy arbitrage.

3.2 Peer-to-peer energy trading

Advanced metering infrastructure and high penetration of distributed energy sources enable direct energy trading between prosumers. This localised energy trading is referred to as peer-to-peer energy trading, transactive energy, or community self-consumption [50]. Exchanging energy between prosumers results in both bidirectional capital and energy flows within the local energy systems as shown in **Figure 7**. Managing for these bidirectional flows not only requires the localised energy markets and trading mechanisms, but also relies on prosumers to coordinate their distributed energy sources.

Energy storage will improve prosumers ability to participate in peer-to-peer energy trading. Energy storage systems give individual prosumers freedom to make strategic energy trading and battery scheduling decisions to achieve their objectives, e.g. reducing electricity bills or maximising revenue from energy trading. For distribution systems, energy storage can offset the energy imports from the utility grid, in particularly during the peak demand periods. For the whole energy systems.

Tariff	Rate [pence/kWh]	Energy Consumption [kWh]	Charge [pence]
Peak	26	8	208
Off-Peak	10	5	50

Table 4.
Example of energy charge of a consumer under peak and off-peak tariffs.

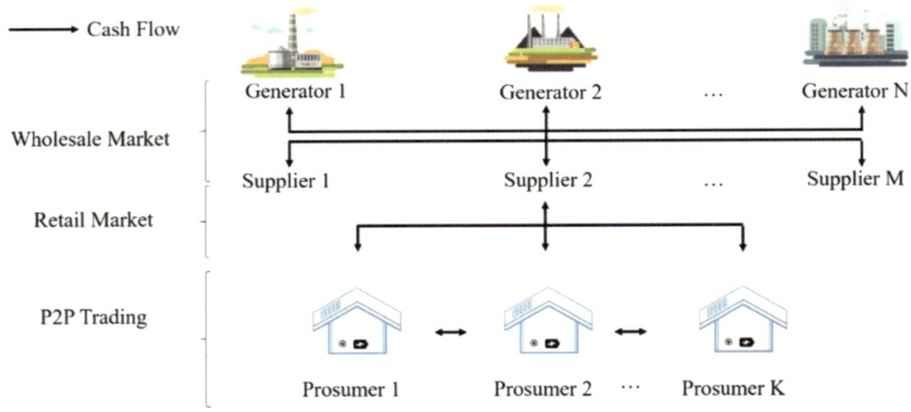

Figure 7.
Schematic illustration of the bidirectional cash flow when a prosumer participates in peer-to-peer energy trading.

Optimally combining energy storage systems with peer-to-peer energy trading can push down market clearing prices and facilitate both the local energy balance and power system energy balance.

3.3 Ancillary services

Local energy systems can participate in a range of ancillary services markets through an aggregator. An aggregator is a company that combines the changes in consumer demand or generation and sells them in markets. The surplus on-site generation of an ensemble of prosumers can be aggregated and stored in local energy systems and later delivered to the utility grid as a service.

There is a wide range of ancillary service with various time scales, e.g. from months ahead to seconds, and multiple functions, e.g. frequency response, energy reserve, voltage support, and demand response [51]. The applications of energy storage in supporting each function of ancillary services are described as follows:

Frequency Response: Frequency response is an action to stabilise the frequency, once power systems encounter sudden loss of generation or load [51]. Energy storage systems play a significant role for regulating the frequency by absorbing or releasing power in response to the deviation from nominal frequency and imbalances between supply and demand.

Energy Reserve: Energy reserve is additional capacity which is online but unloaded to rapidly compensate outages of generation or transmission [52]. The energy storage system can provide this additional reserve capacity without the need for investing an new generation capacities.

Voltage Support: The integration of distributed renewable energy sources, e.g. solar PV, in low-voltage distribution grids would affect the grid voltage profiles during the high generation and low demand periods. The energy storage provides a solution to ensuring the voltage quality requirement while accommodating more distributed generation capacity. The research in [53] suggested the potential of incorporating the storage with reactive power methods to avoid grid reinforcement and active power curtailment.

Demand Response: Demand response refers to reshaping consumption behaviours in response to the dynamic retail electrical energy pricing signals. Instead of curtailing or shifting away loads during high-price periods, a consumer can release stored energy to reduce electricity bills.

4. Design, operation and control of electricity storage systems

4.1 Design considerations

There are many factors to consider for the design of an electricity storage system for a local energy system. Some of those design considerations are:

- Power rating
- Energy capacity
- Physical size
- Response time
- Capital cost
- Lifetime
- O&M complexity and cost

For some electricity storage applications, there may be a specific power rating and energy capacity requirement. When this is the case, this will dictate the design of the electricity storage system to ensure the requirements are met.

For most applications, further analysis must be carried out to maximise benefits from electricity storage. In most cases, the focus is on selecting the optimal power rating and energy capacity for each case. To do this, an in-depth financial analysis must be carried out on a case-by-case basis. The technologies that show promise for local energy system applications have high energy density and fast response times, reducing the need for analysis of physical size and response time.

4.2 Electricity storage operation within a local energy system

Optimal power rating and energy capacity are closely linked to the electricity storage application. Specifically, the way in which the operator intends to use the storage system to benefit themselves. There are any number of operating objectives that the owner of an energy storage system may choose to prioritise. Broadly, these can be assigned to two categories:

- Maximising self-consumption
- Maximising profit by trading electricity

The energy consumed on-site, relative to the total energy generated on-site, is referred to as self-consumption [54]. Energy storage systems, typically batteries, are used as a method to increase self-consumption [54]. This is done by storing surplus electricity from local generation for use at another time, when generation is not enough to meet demand. Self-consumption is particularly attractive due to the difference in import and export electricity prices. The difference between import and export prices is due to some additional charges that apply to import prices. Transmission and distribution network charges recover the cost of building and maintaining all assets in the electricity transmission and distribution networks.

These are charges that are paid by consumers when they purchase electricity from the grid, effectively increasing their electricity import price. These costs are not recouped when consumers sell electricity back to the grid.

Additional charges that are included in a consumers electricity bill are the supplier operating cost, supplier profit and in some circumstances social and environmental costs. All these costs are added to the basic electricity price to give the customer purchase price. The customer sell price however, consists of only the basic price of electricity. Therefore, the purchase price is higher than the sell price. So, from the local energy system perspective, self-consumption of renewable generation will reduce costs by minimising purchased electricity, effectively avoiding extra charges.

Historically, trading has referred to the trading of energy in the wholesale markets by generators, retailers and large industrial consumers [55]. More recently, smaller generators and consumers can also trade electricity between themselves locally or in the wholesale market through an aggregator [55]. Therefore, energy trading is the buying/selling of energy, whether that be in national or local markets. This allows the maximisation of financial benefits for any grid connected consumer that can control their power exchange with the grid. Electricity storage can give local energy systems the flexibility to capitalise on energy trading to maximise their profits. Energy can be stored while the electricity price is low, and discharged when the price is high, this is referred to as price arbitrage.

4.3 Case study: optimal sizing of electricity storage for local energy systems

When local energy systems are considering installing an electricity storage system, often the most important decision is what should be the power rating and energy capacity of the battery. Among other reasons, this is due to the capital cost of electricity storage being closely related to the power rating and energy capacity. The electricity storage system must be large enough to deliver tangible benefits to the owner and operator. However, the larger the power rating and energy capacity, the higher the investment cost will be. Therefore, a balance must be found between the investment cost and the benefits the storage system can provide. The total lifetime costs of a battery storage system are [56, 57]:

- Total investment or 'capital' cost (£/kWh and £/kW). This cost is paid when the battery storage system is purchased and is determined by the power rating and energy capacity prices.

- Operating cost savings (£/year). These are the operational cost savings that battery storage systems provide over their lifetime.

- Maintenance cost (£/kW/no. of years). Some battery storage systems may require maintenance, the cost of which is determined by the power rating and energy capacity and is paid when necessary. This can be after a specific time frame (for example, every one to five years) or irregularly.

- Replacement and/or disposal. At the battery storage systems end-of-life, it must be either repurposed for a second life application or disposed of. In either case, the battery must be replaced if flexibility is still desired.

Operational optimisation is a method commonly used to calculate the operating cost of an energy system with a given battery storage size [56]. The following example describes how this method works and applies it to a case study to demonstrate its efficacy. In addition, the results of the operational optimisation are used in a net present value (NPV) calculation to assess the optimal battery storage size. As they have shown the most promise for applications in local energy system storage, electrochemical battery storage is used in this example. Specifically, this example considers a lithium-ion battery storage system.

The objectives of this section are to:

1. Describe a methodology for evaluating the optimal power rating and energy capacity of a battery storage system in a local energy system with on-site renewable generation and load, as well as a bi-directional connection to the grid.

2. Apply the methodology to a suitable case study to demonstrate the process and ensure repeatability of the case study.

3. Present a comparison of net present values for various power ratings and energy capacities of the battery storage system and identify the optimal combination.

4.3.1 Methodology

A common method for assessing the optimal size of a battery storage system is to formulate an optimisation problem, using linear programming to calculate the operating costs with a battery storage system. The operating costs can then be compared with capital costs to estimate the optimal battery size. The following methodology is a general formulation to assess optimal sizing of a battery storage system for any local energy system with on-site renewable generation and load.

The power rating and energy capacity are defined inputs, requiring a range of combinations for a comparison. The capital cost is found using the inputs for power rating and energy capacity and their defined prices. Then, the operational optimisation determines the cost savings from operating the battery storage system for a period of time (usually one year). In this case, NPV is used to determine the efficacy of the investment for reducing operating cost.

The operating cost consists of any fuel from on-site generation and the cost of power exchange with the grid. For this general formulation, one battery lifetime is considered, therefore, no replacement/disposal costs are included. The capital cost is calculated as follows:

$$I^T = I^P + I^E = \psi^P P^{max} + \psi^E E^{max}. \tag{1}$$

Where, I^T is the total capital cost, I^P is the power rating capital cost and I^E is the energy capacity capital cost. The two latter values are found by multiplying the power (ψ^P) and energy (ψ^E) prices with the power rating (P^{max}) and energy capacity (E^{max}) values. The NPV was calculated using the following equation:

$$NPV = -I^T + \sum_{y=1}^{Y} \left(\frac{NCS_y}{(1+i)^y} \right). \tag{2}$$

Where, y is the year ($y = 1, \dots, Y$), i is the discount rate and NCS_y is the net cost savings each year. The net cost savings (NCSs) is the operating cost with no battery storage (O^{No}), minus the operating (O_y) and the maintenance (M) cost with battery storage. Effectively, this is the cost savings made by including the battery storage system, shown in (3).

$$NCS_y = O^{No} - O_y - M \tag{3}$$

Where, M is the maintenance cost required for specific battery installations. This could be frequent (every year) or infrequent (every 5 years). The operating cost with no battery storage and yearly operating cost with a battery storage system are found using the operational optimisation. The objective of the operational optimisation is to minimise the total operating cost for the local energy system.

$$\text{Min} \sum_{y=1}^{Y} (O_y) \tag{4}$$

Where,

$$O_y = \sum_{t=1}^{T} \tau \left(\psi_t^{im} P_{t,y}^{im} - \psi_t^{ex} P_{t,y}^{ex} \right) \tag{5}$$

In (4), the total operating cost for all years is minimised. The operating cost for each year is determined by (5), as the difference between the cost of importing electricity and the revenue from exporting electricity. The cost of importing electricity is found by multiplying the import price (ψ_t^{im}) by the import power ($P_{t,y}^{im}$) and the time interval (τ) in each time step. The equivalent can be done for the export price (ψ_t^{ex}) and export power ($P_{t,y}^{ex}$) to give the revenue from exporting. The power terms are multiplied by τ to convert from power to energy, as the import and export prices are per unit energy (£/kWh). The index t is the time step ($t = 1, \dots, T$) and T is the total number of time steps, or the time horizon. In this case, the local energy system is considered as a price taker in a retail market, simply agreeing a fixed or time-of-use contract with a supplier.

For each time step within each year, the power balance of the local energy system must be met. This is shown in (6).

$$P_{t,y}^{im} + P_{t,y}^{dis} + P_t^{Ren} = P_{t,y}^{ex} + P_{t,y}^{ch} + P_t^{L} \tag{6}$$

Where the import power, battery discharge power ($P_{t,y}^{dis}$) and on-site renewable generation (P_t^{Ren}) are equal to the export power, battery charging power ($P_{t,y}^{ch}$) and local load (P_t^{L}). Eq. (6) ensures that all power flows are balanced, and that local demand is met. This example does not consider the curtailment of on-site renewable generation. This is justified as no grid constraints are considered and because for any positive export price, the local energy system will export generation, rather than curtail it. The battery storage system constraints are shown in (7)–(11).

$$0 \le P_{t,y}^{ch} \le P^{max} \tag{7}$$

$$0 \le P_{t,y}^{dis} \le P^{max} \tag{8}$$

$$E^{min} \le E_{t,y} \le E_y^{deg} \tag{9}$$

$$E_y^{deg} = E^{max} - \delta y E^{max} \tag{10}$$

$$E_{t,y} = E_{t-1,y} + \tau \left(\eta^{ch} P_{t,y}^{ch} - \frac{P_{t,y}^{dis}}{\eta^{dis}} \right) \tag{11}$$

Eqs. (7) and (8) are the charging and discharging constraints and ensure that the charging/discharging power ($P_{t,y}^{ch}/P_{t,y}^{dis}$) is always within the batteries rated power (P^{max}). Eq. (9) ensures that the energy stored in the battery ($E_{t,y}$) is within the available battery capacity (E_y^{deg}). Where, E_y^{deg} is the available energy capacity of the battery, after degradation has occurred. The available energy capacity is defined in (10), where δ is the percentage degradation per year, y is the year and E^{max} is the battery energy capacity. Eq. (11) is the energy balance equation, which defines how the energy stored in the battery changes from one time step to the next. Both charging and discharge are subject to efficiency losses, which are accounted for with η^{ch} and η^{dis}. In this formulation, the power rating and energy capacity are defined inputs. The model is run with P^{max} and E^{max} as fixed values. These are manually varied, and the optimisation repeatedly run to give operating costs for different sized battery storage systems. The operational cost with no battery storage (O^{No}) is found by inputting power rating and energy capacity as zero.

Grid connection limits are not considered in this formulation. However, these can be included as additional constraints, limiting $P_{t,y}^{im}$ and $P_{t,y}^{ex}$ to a maximum value.

4.3.2 Case study definition

The case study was designed to represent a general local energy system with on-site renewable generation, local load and a connection to the grid. The local energy system needs to install an optimally sized battery storage system to reduce their electricity costs. The analysis was carried out at half hourly time intervals (τ = 0.5 hour), over a full year of operation. The battery's power rating and energy capacity were varied to assess the NPV for a range of combinations. In this case study, intervals of 5 kWh were used to assess the energy capacity and 0.5 kW to assess the power rating. Smaller intervals in power rating and energy capacity would lead to a more accurate outcome but would increase the computing time. Lithium-ion energy capacity prices are projected to fall to as low as £52/kWh by 2030 [58]. Power rating and energy capacity costs similar to this were used for this case study and are shown in **Table 5**. Modern small-scale lithium-ion battery storage systems require no maintenance over their lifetime [59]. Therefore, the maintenance cost (M), was assumed to be £0/kW/year. The interest rate used for the NPV calculation was 10%. Remaining battery costs and characteristics are defined in **Table 5**.

Characteristic	Value
Power rating capital cost, ψ^P (£/kW)	60
Energy capacity capital cost, ψ^E (£/kWh)	60
Maintenance cost (£/kW/year)	0
Charging efficiency, η^{ch} (%)	94.9
Discharging efficiency, η^{dis} (%)	94.9
Degradation, δ (% of energy capacity/year)	2
Lifetime, Y (years)	10

Table 5.
Battery cost and technical characteristics.

The battery round trip efficiency was assumed to be 90%. Therefore, the charging/discharging efficiencies were the square root of 90%. Battery degradation was included as a reduction of energy capacity that was applied to each year, for the duration of the battery lifetime. A degradation of 2% per year, resulted in 80% of the original energy capacity after the lifetime of the battery (10 years). In some cases, battery system operators will limit the maximum depth of discharge to reduce degradation and extend the lifetime of the battery. In this case, no depth of discharge limit was considered ($E^{min} = 0$).

The local energy systems renewable generation was assumed to be from 30 kW of solar photovoltaic (PV) panels. Publicly available solar generation data was used as the input for PV generation [60]. Additionally, the input load data was publicly available non-domestic load averages for weekdays, Saturdays and Sundays for five sections of the year [61]. Both PV generation and load data are presented in **Figure 8**.

The import and export prices were also defined input data. The import contract was a non-domestic day/night tariff offered by Octopus Energy, where night was considered as the seven hours between 00:00 and 07:00 [62, 63]. The export contract was a fixed tariff offered by Octopus Energy [64]. The import and export tariffs are shown in **Figure 9** for a 24-hour period. The import and export prices were the same every day of the year.

4.3.3 Results

As the lowest import price does not go below the highest export price in this example, energy trading for price arbitrage was not possible. Therefore, the local

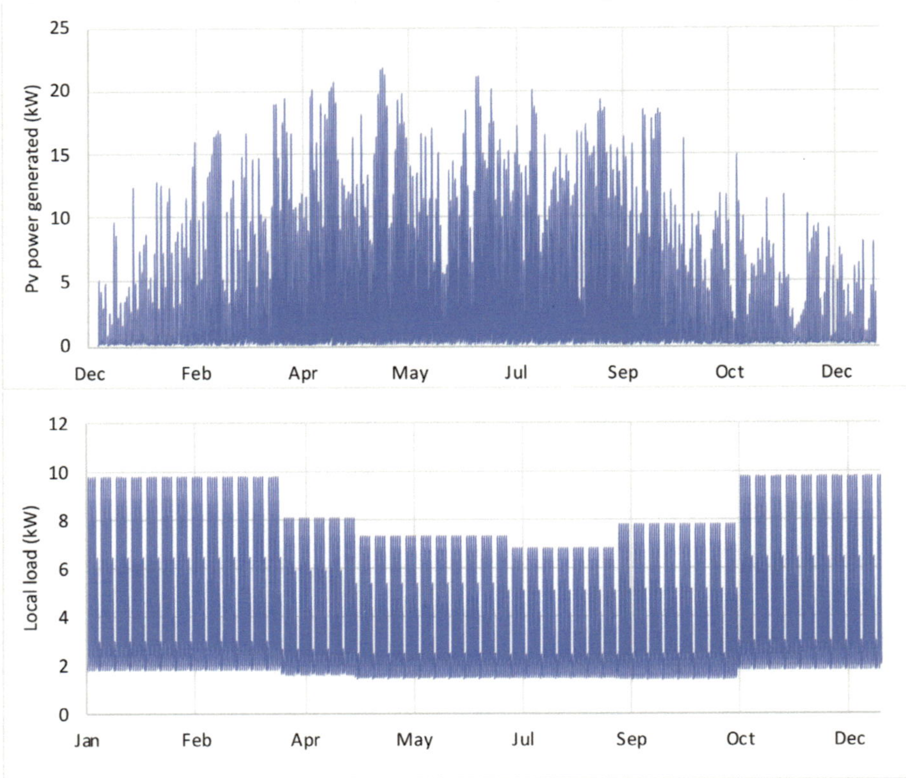

Figure 8.
On-site PV generation and local load profiles for all time steps over a year.

energy system operating objective was to maximise self-consumption of renewable generation, to minimise electricity purchased from the grid.

The local energy systems operating cost for one year with no battery storage system was £2039.14. The operating cost reduced as the battery power rating and energy capacity were increased. However, as the power rating and energy capacity were increased, the capital cost increased. To maximise the benefit of investing in battery storage, a balance must be found between the capital cost and the operating cost savings. NPV was used to assess the lifetime value of a range of power rating and energy capacity combinations. The results are shown in **Figure 10**.

The results in **Figure 10** show that the NPV was impacted by both the power rating and energy capacity. In addition, for each battery energy capacity, there was an optimal power rating. As an example, for a battery energy capacity of 10 kWh, the optimal power rating was approximately 2 kW. Furthermore, the result showed a clear optimal energy capacity of 20 kWh, with an optimal power rating of

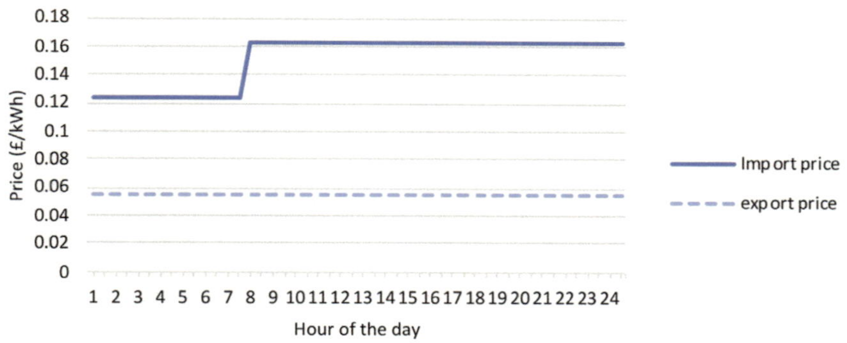

Figure 9.
Import and export prices in the local energy systems 12-month contract with a retailer.

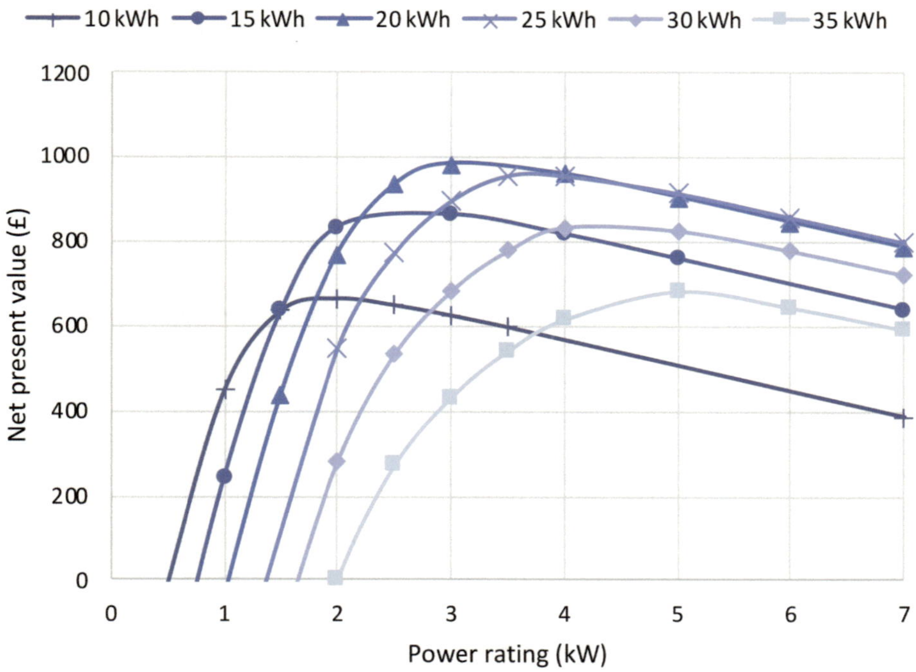

Figure 10.
NPV for a range of power rating and energy capacity combinations.

approximately 3 kW. This result was taken from the highest peak in **Figure 10**, representing the highest value of NPV. This result demonstrates the need to carry out a detailed financial analysis to determine the most suitable power rating and energy capacity for a local energy system.

These results are significantly affected by the assumed input data. In particular, the power rating and energy capacity capital costs. To ensure reliability of result, the input data, especially the prices, must be accurate.

4.3.4 Conclusion

In conclusion, this case study was carried out to describe a methodology for optimal sizing of battery storage for a local energy system. Firstly, a methodology based on operational optimisation and net present value was used to assess the value of adding different combinations of battery power ratings and energy capacities. Then, a case study was defined to demonstrate the practical use of the methodology. The results show clear optimal power ratings for each energy capacity, shown as a peak in net present value. The optimal battery power rating and energy capacity for the case study were 3 kW and 20 kWh, approximately. The result shows that this method can assess the optimal size of a battery storage system for a local energy system. Furthermore, the method can be applied to a wide variety of local energy system configurations, enabling the method to be tailored to specific cases. Finally, the accuracy of the outcome heavily relies on the accuracy of the input data. Therefore, the more detailed and accurate the input data, the more reliable the result.

5. Conclusions

The power system transition from large scale power generation to decentralised integration of distributed energy sources is creating an increasing need for flexibility. With higher penetrations of intermittent wind and solar generation and a decrease in dispatchable fossil-based power plants, flexibility is required to reduce curtailment of renewable generation and ensure a stable frequency by balancing supply and demand. Electricity storage systems can save excess energy during high renewable generation and release it when there is high demand or low renewable generation.

Many electricity storage technologies exist today that store energy in mechanical, chemical, electrical and magnetic mediums. Technologies such as pumped hydro storage and compressed air energy storage have geographical requirements that are generally less suited to local energy systems. Flywheels, capacitors and supercapacitors have relatively low energy storage capacity, limiting their use in a local energy system. In comparison, cryogenic energy storage has high energy density but is limited by its low efficiency of approximately 50%.

The most promising technology for local energy system electricity storage is electrochemical. In particular, lithium-ion battery storage systems have high energy density, longevity, high efficiency and rapid response. Alongside disadvantages such as environmental impact of lithium production and recycling, lithium-ion batteries have a high capital cost. However, increasing demand for large batteries for electric vehicles has driven down the price by 89% over the last 10 years. This, along with improved remanufacturing, refurbishing and recycling technologies has put lithium-ion batteries at the forefront of electricity storage in recent years.

Applications of electricity storage for local energy systems include self-consumption, energy trading and providing services to the utility grid. Local energy systems can strategically store energy when the electricity price is low and release it when the price is high. This is called energy arbitrage and is a common strategy employed by owners of electricity storage systems. An alternative energy trading strategy is peer-to-peer trading, where energy is traded between prosumers and consumers in bidirectional agreements. Finally, ancillary services are a group of flexibility markets that local energy systems can participate in via an aggregator. There is a variety of services including frequency response, energy reserve, voltage support and demand response.

The design and operation of an electricity storage system for a local energy system is unique. Several considerations must be made, including power rating, energy capacity, physical size, response time, capital cost, lifetime and operation and maintenance cost. A balance must be struck between each characteristic to ensure the most beneficial design for the local energy system. A case study provides a methodology for obtaining the optimal size of a battery storage system. Specifically designed for repeatability, the energy generation and demand can be adapted to fit any local energy system configuration. The optimal power rating and energy capacity were determined and presented in the results.

Nomenclature

Symbol	Description
Sets	
t	Time step index
y	Number of years index
Input parameters	
E^{max}	Battery energy capacity (kWh)
E^{min}	Minimum battery state of charge (kWh)
O^{No}	Local energy system operating cost with no battery storage (%)
P^{max}	Battery power rating (kW)
P_t^L	Local energy system on-site load (kW)
P_t^{Ren}	Local energy system renewable power generation (kW)
η^{ch}	Battery charging efficiency (%)
η^{dis}	Battery discharging efficiency (%)
ψ^E	Energy capacity price (£/kWh)
ψ^P	Power rating price (£/kW)
ψ_t^{ex}	Export price (£/kWh)
ψ_t^{im}	Import price (£/kWh)
i	Interest rate (%)
T	Total number of time steps
Y	Total number of years
δ	Rate of degradation (%)
τ	Time interval (0.5 hours)
Variables	
$E_{t,y}$	Energy stored in battery (kWh)
E_y^{deg}	Energy capacity available, after degradation (kWh)
I^E	Battery energy capital cost (£)

I^P	Battery power capital cost (£)
I^T	Total capital cost (£)
M	Maintenance cost (£)
NCS_y	Net cost savings for each operating year (£)
NPV	Net present value (£)
O_y	Operating cost with battery storage (£)
$P^{ch}_{t,y}$	Battery charging power (kW)
$P^{dis}_{t,y}$	Battery discharge power (kW)
$P^{ex}_{t,y}$	Export power (kWh)
$P^{im}_{t,y}$	Import power (kW)

Author details

William Seward*, Weiqi Hua and Meysam Qadrdan
Cardiff University, Cardiff, United Kingdom

*Address all correspondence to: sewardwg@cardiff.ac.uk

References

[1] Renewable Energy Foundation, "Balancing Mechanism Wind Farm Constraint Payments," 2021.

[2] Committe on Climate Change (CCC), "Net Zero - The UK's contribution to stopping global warming," 2019.

[3] National Grid ESO, "Future Energy Scenarios," Warwick, UK, 2020.

[4] A. Ajanovic, A. Hiesl, and R. Haas, "On the role of storage for electricity in smart energy systems," *Energy*, vol. 200, 2020.

[5] C. K. Das, O. Bass, G. Kothapalli, T. S. Mahmoud, and D. Habibi, "Overview of energy storage systems in distribution networks: Placement, sizing, operation, and power quality," *Renew. Sustain. Energy Rev.*, vol. 91, no. November 2016, pp. 1205–1230, 2018.

[6] World Energy Council, "Five Steps to Energy Storage," London, 2020.

[7] S. Kalaiselvam and R. Parameshwaran, "Chapter 2 - Energy Storage," in *Thermal Energy Storage Technologies for Sustainability: Systems Design, Assessment and Applications*, Elsevier Science, 2014, pp. 21–56.

[8] Q. Yu *et al.*, "Chapter Eight - Cryogenic Energy Storage and Its Integration With Nuclear Power Generation for Load Shift," H. Bindra and S. B. T.-S. and H. of N. E. Revankar, Eds. Academic Press, 2019, pp. 249–273.

[9] H. Chen, Y. Xu, C. Liu, F. He, and S. Hu, "Chapter 24 - Storing Energy in China—An Overview," T. M. B. T.-S. E. Letcher, Ed. Oxford: Elsevier, 2016, pp. 509–527.

[10] P. Breeze, "Chapter 10 - Power System Energy Storage Technologies," P. B. T.-P. G. T. (Third E. Breeze, Ed. Newnes, 2019, pp. 219–249.

[11] P. Breeze, "Chapter 4 - Large-Scale Batteries," P. B. T.-P. S. E. S. T. Breeze, Ed. Academic Press, 2018, pp. 33–45.

[12] D. Steward, G. Saur, M. Penev, and T. Ramsden, "Lifecycle cost analysis of hydrogen versus other technologies for electrical energy storage." pp. 59–120, 2011.

[13] Ã. Cunha, J. Martins, N. Rodrigues, and F. P. Brito, "Vanadium redox flow batteries: a technology review," *International journal of energy research*, vol. 39, no. 7. [New York, NY] :, pp. 889–918, 2015.

[14] X. Luo, J. Wang, M. Dooner, and J. Clarke, "Overview of current development in electrical energy storage technologies and the application potential in power system operation," *Appl. Energy*, vol. 137, pp. 511–536, 2015.

[15] M. Dooner and J. Wang, "Chapter 14 - Compressed-Air Energy Storage," in *Future Energy*, T. M. B. T.-F. E. (Third E. Letcher, Ed. Elsevier, 2020, pp. 279–312.

[16] M. E. Amiryar and K. R. Pullen, "A review of flywheel energy storage system technologies and their applications," *Appl. Sci.*, vol. 7, no. 3, p. 286, 2017.

[17] H. Chen, T. N. Cong, W. Yang, C. Tan, Y. Li, and Y. Ding, "Progress in electrical energy storage system: A critical review," *Prog. Nat. Sci.*, vol. 19, no. 3, pp. 291–312, 2009.

[18] T. M. Masaud, K. Lee, and P. K. Sen, "An overview of enrgy storage technologies in electric power systems: What is the future?," *North Am. Power Symp. 2010, NAPS 2010*, 2010.

[19] F. Rafik, H. Gualous, R. Gallay, A. Crausaz, and A. Berthon, "Frequency, thermal and voltage supercapacitor characterization and modeling," *J. Power Sources*, vol. 165, no. 2, pp. 928–934, 2007.

[20] J. M. Baptista, J. S. Sagu, U. W. KG, and K. Lobato, "State-of-the-art materials for high power and high energy supercapacitors: Performance metrics and obstacles for the transition from lab to industrial scale–A critical approach," *Chem. Eng. J.*, vol. 374, pp. 1153–1179, 2019.

[21] Y. Ding, L. Tong, P. Zhang, Y. Li, J. Radcliffe, and L. Wang, "Chapter 9 - Liquid Air Energy Storage," in *Storing Energy*, T. M. B. T.-S. E. Letcher, Ed. Oxford: Elsevier, 2016, pp. 167–181.

[22] R. Moradi and K. M. Groth, "Hydrogen storage and delivery: Review of the state of the art technologies and risk and reliability analysis," *Int. J. Hydrogen Energy*, vol. 44, no. 23, pp. 12254–12269, 2019.

[23] B. Zakeri and S. Syri, "Electrical energy storage systems: A comparative life cycle cost analysis," *Renew. Sustain. Energy Rev.*, vol. 42, pp. 569–596, 2015.

[24] S. Koohi-Fayegh and M. A. Rosen, "A review of energy storage types, applications and recent developments," *J. Energy Storage*, vol. 27, no. July 2019, p. 101047, 2020.

[25] X. Zeng *et al.*, "Commercialization of Lithium Battery Technologies for Electric Vehicles," *Adv. Energy Mater.*, vol. 9, no. 27, pp. 1–25, 2019.

[26] A. Sánchez Muñoz, M. Garcia, M. Gerlich Reviewer, and T. Rautiainen, "Overview of storage technologies," no. 645963, 2016.

[27] K. Mongird *et al.*, "Energy Storage Technology and Cost Characterization Report | Department of Energy," no. July, 2019.

[28] P. Nikolaidis and A. Poullikkas, "Cost metrics of electrical energy storage technologies in potential power system operations," *Sustain. Energy Technol. Assessments*, vol. 25, no. December 2017, pp. 43–59, 2018.

[29] The Faraday Institution, "High-Energy Battery Technologies," 2020.

[30] T. Aquino, M. Roling, C. Baker, and L. Rowland, "Battery Energy Storage Technology Assessment," 2017.

[31] Energy Sector Management Assistance Program and Climate Smart Mining Initiative, "Reuse and Recycling: Environmental Sustainability of Lithium-Ion Battery Energy Storage Systems," Washington, 2020.

[32] T. Chen *et al.*, "Applications of Lithium-Ion Batteries in Grid-Scale Energy Storage Systems," *Trans. Tianjin Univ.*, vol. 26, no. 3, pp. 208–217, 2020.

[33] M. Armand *et al.*, "Lithium-ion batteries – Current state of the art and anticipated developments," *J. Power Sources*, vol. 479, no. July, p. 228708, 2020.

[34] T. Kim, W. Song, D.-Y. Son, L. K. Ono, and Y. Qi, "Lithium-ion batteries: outlook on present, future, and hybridized technologies," *J. Mater. Chem. A*, vol. 7, no. 7, pp. 2942–2964, 2019.

[35] G. Zubi, R. Dufo-López, M. Carvalho, and G. Pasaoglu, "The lithium-ion battery: State of the art and future perspectives," *Renew. Sustain. Energy Rev.*, vol. 89, no. March, pp. 292–308, 2018.

[36] On Location, "Electricity Storage Technology Review," 2020.

[37] X. Han *et al.*, "A review on the key issues of the lithium ion battery degradation among the whole life

cycle," *eTransportation*, vol. 1, p. 100005, 2019.

[38] G. Saldaña, J. I. S. Martín, I. Zamora, F. J. Asensio, and O. Oñederra, "Analysis of the current electric battery models for electric vehicle simulation," *Energies*, vol. 12, no. 14, 2019.

[39] M. Chen *et al.*, "Recycling End-of-Life Electric Vehicle Lithium-Ion Batteries," *Joule*, vol. 3, no. 11. pp. 2622–2646, 2019.

[40] D. C. R. Espinosa and M. B. Mansur, "17 - Recycling batteries," in *Woodhead Publishing Series in Electronic and Optical Materials*, V. Goodship and A. B. T.-W. E. and E. E. (WEEE) H. Stevels, Eds. Woodhead Publishing, 2012, pp. 365–384.

[41] A. Mayyas, D. Steward, and M. Mann, "The case for recycling: Overview and challenges in the material supply chain for automotive li-ion batteries," *Sustain. Mater. Technol.*, vol. 19, p. e00087, 2019.

[42] CleanTechnica, "Global Plugin Vehicle Sales Up 43% In 2020, European Sales Up 137%," 2021. [Online]. Available: https://cleantechnica.com/2021/02/08/global-plugin-vehicle-sales-up-43-in-2020-european-sales-up-137/. [Accessed: 10-May-2021].

[43] BloombergNEF, "Battery Pack Prices Cited Below $100/kWh for the First Time in 2020, While Market Average Sits at $137/kWh," 2020. [Online]. Available: https://about.bnef.com/blog/battery-pack-prices-cited-below-100-kwh-for-the-first-time-in-2020-while-market-average-sits-at-137-kwh/. [Accessed: 01-Feb-2021].

[44] N. Bullard, "This Is the Dawning of the Age of the Battery," *Bloomberg Green*, 2020. [Online]. Available: https://www.bloomberg.com/news/articles/2020-12-17/this-is-the-dawning-of-the-age-of-the-battery. [Accessed: 01-Feb-2021].

[45] Ioannis Tsiropoulos, Dalius Tarvydas, and Natalia Lebedeva, *Li-ion batteries for mobility and stationary storage applications - Scenarios for costs and market growth* . 2018.

[46] International Renewable Energy Agency (IRENA), "Behind the Meter Batteries," 2019.

[47] H. Lund and E. Münster, "Integrated energy systems and local energy markets," *Energy Policy*, vol. 34, no. 10, pp. 1152–1160, 2006.

[48] S. A. El-Batawy and W. G. Morsi, "Optimal design of community battery energy storage systems with prosumers owning electric vehicles," *IEEE Trans. Ind. Informatics*, vol. 14, no. 5, pp. 1920–1931, 2017.

[49] M. S. Javadi, M. Gough, M. Lotfi, A. E. Nezhad, S. F. Santos, and J. P. S. Catalão, "Optimal self-scheduling of home energy management system in the presence of photovoltaic power generation and batteries," *Energy*, vol. 210, p. 118568, 2020.

[50] R. B. Melton and M. Knight, "Gridwise transactive energy framework version 1," Richland, US, 2015.

[51] D. Chakravorty, B. Chaudhuri, and S. Y. R. Hui, "Rapid frequency response from smart loads in Great Britain power system," *IEEE Trans. Smart Grid*, vol. 8, no. 5, pp. 2160–2169, 2016.

[52] M. Bozorg, F. Sossan, J.-Y. Le Boudec, and M. Paolone, "Influencing the bulk power system reserve by dispatching power distribution networks using local energy storage," *Electr. Power Syst. Res.*, vol. 163, pp. 270–279, 2018.

[53] F. Marra, Y. T. Fawzy, T. Bülo, and B. Blažic, "Energy storage options for voltage support in low-voltage grids with high penetration of photovoltaic,"

in *2012 3rd IEEE PES Innovative Smart Grid Technologies Europe (ISGT Europe)*, 2012, pp. 1–7.

[54] R. Luthander, J. Widén, D. Nilsson, and J. Palm, "Photovoltaic self-consumption in buildings: A review," *Appl. Energy*, vol. 142, pp. 80–94, 2015.

[55] M. Khorasany, Y. Mishra, and G. Ledwich, "Market framework for local energy trading: A review of potential designs and market clearing approaches," *IET Gener. Transm. Distrib.*, vol. 12, no. 22, pp. 5899–5908, 2018.

[56] D. W. Gao, "Chapter 5 - Sizing of Energy Storage Systems for Microgrids," in *Energy Storage for Sustainable Microgrid*, Elsevier Science & Technology, 2015.

[57] G. Carpinelli, F. Mottola, and D. Proto, "Probabilistic sizing of battery energy storage when time-of-use pricing is applied," *Electr. Power Syst. Res.*, vol. 141, pp. 73–83, 2016.

[58] J. Gerdes, "Reducing battery cost is essential for a clean energy future," *Energy Monitor*, 2020. [Online]. Available: https://energymonitor.ai/technology/energy-storage/reducing-battery-cost-is-essential-for-a-clean-energy-future. [Accessed: 27-May-2021].

[59] Tesla, "Powerwall 2 AC: Owner's Manual," *Powerwall*, 2018. [Online]. Available: https://www.tesla.com/sites/default/files/pdfs/powerwall/powerwall_2_ac_owners_manual.pdf. [Accessed: 27-May-2021].

[60] The University of Sheffield, "PV_Live," *Sheffield Solar*, 2021. [Online]. Available: https://www.solar.sheffield.ac.uk/pvlive/. [Accessed: 24-May-2021].

[61] UK Energy Research Council (UKERC), "Electricity user load profiles by profile class," *UKERC Energy Data Centre: Data Catalogue*, 2021. [Online]. Available: https://ukerc.rl.ac.uk/DC/cgi-bin/edc_search.pl?GoButton=Detail&WantComp=42&&RELATED=1. [Accessed: 24-May-2021].

[62] Octopus Energy, "Octopus green business 12m fixed january 2019," *Our Tariffs*, 2021. [Online]. Available: https://octopus.energy/tariffs/. [Accessed: 24-May-2021].

[63] S. Bridgen, "Octopus Energy claims it will pay you to use electricity - but beware the hype," *MSE News*, 2018. [Online]. Available: https://www.moneysavingexpert.com/news/2018/02/an-energy-firm-claims-its-new-tariff-may-pay-you-to-use-electricity—but-beware-the-hype-/#:~:text=The tariff works in a,day%2C usually 12am to 7am. [Accessed: 24-May-2021].

[64] Octopus Energy, "Introducing Outgoing Octopus," 2021. [Online]. Available: https://octopus.energy/outgoing/. [Accessed: 24-May-2021].

Chapter 6

Local Energy Systems in Iraq: Neighbourhood Diesel Generators and Solar Photovoltaic Generation

Ali Al-Wakeel

Abstract

Iraqis experience interruptions of the public electricity supply of up to 18 hours a day. In response, private entrepreneurs and the Local Provincial Councils (LPCs) have installed an estimated 55,000–80,000 diesel generators, each rated typically between 100 and 500 kVA. The generators supply neighbourhoods through small, isolated distribution networks to operate lighting, fans and small appliances when power is not available from the public supply. A single radial live conductor connects each customer to the generator and payment for the electricity is based on a monthly charge per ampere. The operation and regulation of the neighbourhood diesel generator networks was reviewed through a comprehensive literature survey, site visits and interviews conducted with local operators and assemblers of the generator sets. The electricity is expensive, the generators can only supply small loads, have considerable environmental impact and the unusual single wire distribution practice is potentially hazardous. However, the use of the generators is likely to continue in the absence of any alternative electricity supply. The diesels and networks are poorly regulated and there is scope to enforce existing standards and develop a new standard to address the hazards of the connection practice. The chapter goes on to assess the possibilities of using small photovoltaic systems for power generation in Iraq.

Keywords: diesel generator, informal electricity supply, neighbourhood diesel generators, photovoltaic generation

1. Introduction

Iraq has the 5th largest oil reserves in the world and exported, in October 2020, some 2.88 million barrels of oil per day [1] down from an average of 3.97 million barrels per day in 2019 [2]. The drop in oil exports comes in response to an agreement between worldwide oil producers to cut production and revive the oil market in response to the coronavirus global lockdowns and a collapsing demand for oil [3]. Iraq also has the world's 11th largest reserves of natural gas [4]. However, following four decades of war and international sanctions, the electricity supply system is now in a poor condition and unable to supply the rapidly increasing demand for electricity of a growing population [5].

The electricity infrastructure of Iraq was severely damaged during the First Gulf War in 1991. The sanctions imposed by the United Nations during the early 1990s further reduced electricity supply [6]. In 2003, following the Second Gulf War, the power generated fell from a pre-war value of 5300 MW to 3500 MW whereas the peak demand at that time was estimated to be 6000 MW [7]. Despite the rehabilitation of old power plants and construction of new ones, an annual rate of increase of electrical demand of more than 10% means there is now an estimated deficit of generating capacity of more than 10,000 MW [5].

After the Second Gulf War, the shortage of power led the Iraqi government to encourage the use of neighbourhood diesel generators and novel local distribution networks. Exact details of the numbers of these generators are not available. Reference [8] estimates there are 55,000–80,000 neighbourhood generators while reference [9] reports that the actual number of these generators is between 90,000 – 150,000. These medium sized (100–500 kVA) diesel generators supply 90–95% of households with about 20–30% of their electricity [5, 8]. This unusual community response to electricity shortages by using medium-size diesel generators serving neighbourhoods through a novel distribution network and tariff system is in contrast to some other oil-rich countries with poor public electrical infrastructure where small generators serve only individual consumers.

Over the last three years, encouraged by the falling costs of photovoltaic (PV) modules in international markets, the public have shown growing interest in installing rooftop solar PV systems. These small-sized (1–10 kW) systems are deployed to help residents supplement the public electricity supply and reduce their electricity bills by minimising their dependence upon expensive and polluting neighbourhood generators [10–12]. On the other hand, the Iraqi government has invited independent power producers (IPPs) to develop seven utility-scale PV solar power sites in the range between 30 and 300 MWp with a total power generation capacity of 755 MWp [13]. However, taking into consideration the recent dramatic drop in oil prices, a large deficit in the federal budget and the outbreak of the COVID-19 virus, it is thought to be unlikely that those utility-scale projects will become operational (as planned) by end of 2021 [12, 14].

2. Electricity supplies in Iraq

The Iraq public electricity system is divided into two networks, which have very limited interconnection. The smaller network of around 7000 MW of power generation capacity (in 2019) is owned and operated by the Ministry of Electricity in the Kurdistan Region of Iraq [15]. The larger network of around 27,300 MW of generation capacity, which is the focus of this study, covers Iraq Excluding Kurdistan (IEK) and is owned and operated by the Federal Ministry of Electricity.

The capacity of power generation installed in Iraq (IEK) in 2018 is shown in **Table 1**. It can be seen that the mean generation is considerably less than the installed capacity in spite of the high demand for electricity, indicating power plant is often unavailable. Generation is from gas and steam turbines with some hydropower. The large diesels listed in **Table 1** have capacities of up to 23 MW and are operated by the Federal Ministry of Electricity using heavy fuel oil.

Table 2 lists the types of fuel used in central power plants in 2018. The steam turbines are fuelled mainly by crude oil while most gas turbines are supplied by natural gas. Some gas turbines have been modified to burn crude oil, but these are then de-rated from a nameplate capacity of 2878 MW to a mean generation of 1178 MW.

The Iraqi transmission networks (400 kV in IEK only and 132 kV throughout Iraq) connect the central power plants with load centres [17]. Distribution networks use 33 kV and 11 kV to distribute the power supplied by the transmission network between primary and secondary substations and 0.4/0.23 kV to supply end-users

Power generation technology	All units	Operating units	
	Nameplate capacity (MW)	Mean generation (MW)	Generation capacity factor
Steam turbines	7305	3270	~ 0.448
Gas turbines	15,857	5521	~ 0.348
Large diesels	2327	376	~ 0.162
Hydroelectric turbines	1864	208	~ 0.112
IPPs and Imports	—	3627	—
Total	**27,353**	**13,002**	**0.475**

Table 1.
Nameplate and available capacities of IEK power generation in 2018 [16].

Power generation technology	% of total fuel burnt in generation plants				
	Crude oil	Heavy fuel oil	Light diesel	Gas oil	Natural gas
Steam turbines	75%	12.8%	~ 0%	~ 0%	12.2%
Gas turbines	20.3%	11.4%	2.2%	3.8%	62.3%
Large Diesels	0%	96.6%	0%	3.4%	0%

Table 2.
Type of fuel burnt in IEK central power plants in 2018 [16].

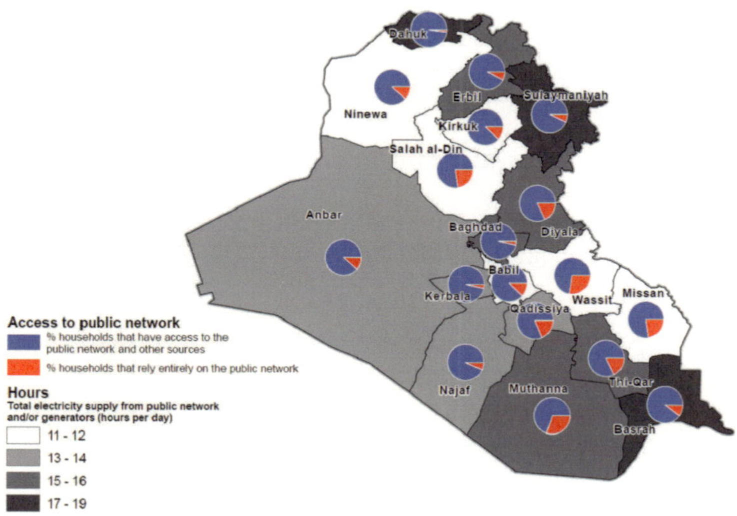

Figure 1.
Source and hours of electricity supplied to residential customers in Iraq, 2011 (Source: [18]).

with electricity. Distribution networks in Iraq are unreliable due to unplanned network growth, shortage of spare parts and lack of maintenance. The absence of effective metering and billing leads to widespread under-collection of revenues [8]. According to the International Energy Agency (IEA) [5], the aggregate technical and non-technical losses in transmission and distribution networks in IEK are between 50 and 60% of the total electrical energy generated and imported, and are among the highest in the world.

The Federal Ministry of Electricity estimated the mean power demand in 2018 to be 22,530 MW but the mean power dispatched to supply this demand was 12,109 MW [16]. For the same year, the IEA estimated the summer peak demand was about 27,000 MW and the unmet demand to be 10,500 MW [5]. The lack of electricity leads to severe hardship in a country where daytime temperatures in summer regularly exceed 45°C and has prompted most households and small businesses to rely on electricity from neighbourhood diesel generators. The source and hours of electricity supplied to residential customers in 2011 are shown in **Figure 1**. This shows that, even at that time, most households supplemented their electricity supply from the public network with a connection to the neighbourhood diesel generators or by using small individual household gasoline generators (or both). This practice of supplementing the public supply remains common. In summer 2020, electricity consumers in IEK received an average of 14–16 hours of electricity per day, with only 6–8 hours provided by the public network [19, 20].

3. Neighbourhood diesel generators in Iraq

In response to this power deficit, private entrepreneurs and the Local Provincial Councils (LPCs) have been encouraged by the government to install medium-sized diesel generators at a neighbourhood level to supplement grid supply and alleviate some power shortages particularly in the peak summer months. These generators are owned and operated either by independent entrepreneurs or by the LPCs. In Baghdad, around 18% of more than 13,000 neighbourhood generators are owned and operated by the LPCs [21]. The generating sets are usually assembled locally from reused truck diesel engines coupled to imported generators, as shown in **Figure 2**. An assembly line of a local assembler of generating sets is shown in **Figure 3**. The control panels are manufactured locally using imported components. The price of a locally assembled 250 kVA generating set is between $8500–10,000 compared to the cost of a UK made imported 220 kVA unit of $18,000 – 19,000. Larger generating sets with capacities up to 2500 kVA are imported as complete units and operated by the LPCs.

The Federal Ministry of Electricity and the LPCs regulate the installation and connection of neighbourhood diesel generators. The 'Regulations of Power

Figure 2.
Locally assembled neighbourhood generator (Source: Author).

Figure 3.
Assembly line of a local manufacturer of neighbourhood diesel generators (Source: Local assembler).

Supply to End Customers' of the Federal Ministry of Electricity [22] requires that neighbourhood diesels are electrically isolated from the public network. Electrical protection must be installed to ensure no current can flow from a generator into the public network. The Federal Ministry of Electricity records the numbers and sites of neighbourhood diesels and they cannot be relocated without obtaining permission from the Ministry and the LPCs. Apart from the technical details given in the 'Regulations of Power Supply to End Customers' and health and safety regulations, all other communications provided by the LPCs and Ministry of Electricity are guidelines only.

The LPCs provide the sites for the neighbourhood diesels, which are typically located on roadside and mid-road pavements, in public parks and near local markets [23]. **Figure 4** shows an example of a neighbourhood generator installation in urban Baghdad. Also, the LPCs define the tariffs used to charge customers and the number of hours that the neighbourhood diesels operate, but these vary between different provinces. Several campaigns by non-governmental organisations and the general public have called for clarification and enforcement of the policies and regulations for operating neighbourhood diesels.

The Ministry of Oil provides the fuel necessary to operate the neighbourhood diesels. The Ministry of Oil defines the amounts of fuel to be supplied, according to the power available from central power plants and anticipated customer electricity demand (**Table 2**). Between 2003 and 2017, fuel was provided initially free-of-charge and later at subsidised rates. After 2017, the diesel fuel was sold to entrepreneurs and LPCs at the regular retail price of 34 US cents per litre (**Table 3**).

Most residential premises in Iraq pay two monthly electricity bills, the first to the Federal Ministry of Electricity and the second to the operator of their local

Figure 4.
Neighbourhood diesel generator installed on a mid-road pavement in Baghdad (Source: Author).

Months	Amount of fuel (litre/kVA)
Summer: May, June, July, August, September	20–35
Spring: March, April	10–15
Autumn: October, November	10–15
Winter: December, January, February	5–10

Table 3.
Amount of fuel per month supplied by the Ministry of Oil to neighbourhood diesels [24].

neighbourhood diesel. Electricity from the public network is charged by energy ($/kWh) while the neighbourhood diesels sell electricity based on the maximum current the customer has chosen ($/Amp).

Table 4 shows the domestic tariffs charged by the Federal Ministry of Electricity for different levels of energy consumption. It has been recognised by the World Bank and the International Energy Agency that these very low tariffs do not promote efficient and rational use of electricity and combined with poor billing and collection, they result in very low cost recovery ratios (approximately 10%) and the electricity sector operating at a loss [5, 25].

The monthly tariffs charged by the neighbourhood diesels are divided into two types (**Table 5**). The standard tariffs of restricted hours are defined by the LPCs and apply to all generators while the premium service that provide 24-hour electricity is offered only by the private entrepreneurs. For the premium service, the entrepreneurs buy additional fuel from the Ministry of Oil at approximately 59 US cents per litre [27].

Customer type	Energy consumed (kWh) per month	Tariff (US ¢/kWh)
Residential	1–1500	0.83
	1501–3000	2.92
	3001–4000	6.67
	over 4001	10.00

All costs in this chapter have been converted from Iraqi dinars to US dollars at a rate of 1200:1.

Table 4.
Electrical energy tariffs charged by the Federal Ministry of Electricity in September 2020 [26].

		Summer	Spring & autumn	Winter	Fuel cost to operators (US ¢/litre)
Standard	*Tariffs (US $/Amp)*	10	7.5	5	34
	Hours of operation per day	10	3–10		
Premium	*Tariffs (US$/Amp)*	21	12.5		59
	Hours of operation per day	24			

Table 5.
Approximate monthly tariffs of the private neighbourhood diesel generators.

	Public network		Neighbourhood diesel generators		
	Approximate monthly		Number of amps chosen (Amp)	Approximate monthly bill (US $)	
Season (months)	Energy consumption (kWh)	Bill (US $)		Standard tariff	Premium tariff
May, June, July, August, September	1740	19.5	7	70	147
March, April	780	6.5	5	37.5	62.5
October, November					
December, January, February	1245	10.4		25	
Annual		154.7		575	1172.5

Table 6.
Approximate monthly energy consumption and bills of a typical residential customer.

There is a considerable difference in the price paid for electricity from the public network and the neighbourhood generators. **Table 6** shows the approximate monthly energy consumption and bills for electricity of a typical residential customer in IEK. The calculation assumes a 24/7 supply of electricity from the public grid with an assumed set of appliances in a typical dwelling. The electrical load is that assumed by the Federal Ministry of Electricity to estimate the consumption of households that are without a functioning meter [22]. An on-line calculator using these assumptions has recently been published by the Ministry of Electricity to help customers estimate their consumption and calculate their bills [28]. **Table 6** contrasts this cost with the charge for a neighbourhood diesel to supply only the essential loads of lighting, fans, evaporative air coolers, white goods and home entertainment systems. It can be seen that the neighbourhood diesels provide a much more expensive service to fewer appliances for reduced hours.

4. Connection of neighbourhood diesels

The circuits used for connecting the neighbourhood diesels are unusual and **Figure 5** shows how the neighbourhood diesel generators are connected using radial private wire distribution circuits of single 2.5 mm^2 or 6 mm^2 copper conductors. Single conductors connect a live phase of the neighbourhood diesels to individual

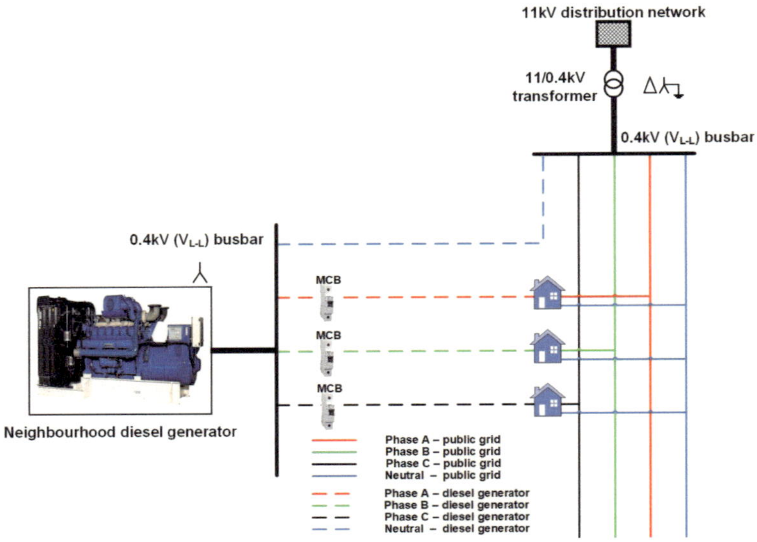

Figure 5.
Simplified diagram of neighbourhood generator connections.

Figure 6.
A miniature circuit breaker board of a neighbourhood diesel generator (Source: Author).

Figure 7.
Informal distribution circuits using redundant utility support insulators (Source: Author).

customer premises. The neutral of the generator is permanently connected to the neutral conductor of the public network. A changeover switch, either automatic or manual, is installed at the customer premises for transfer of the live phase from the public network to the neighbourhood generator when a power outage occurs. There are no clear, enforced regulations for neutral earthing (grounding) but a common practice is not to earth the generator neutral but rely on the earth of the neutral of the 11/0.4 kV distribution transformer.

This unusual neutral connection practice would contravene safety regulations in many countries. There is no means of detecting if the connection between the generator and transformer neutral is lost and, in this condition, ill-defined "floating voltages" will appear at consumers premises. If the single-phase wire to a dwelling is broken, then again there is no means of detecting this and a hazardous voltage may result. There is considerable anecdotal evidence of fires being caused by faults on these circuits and electrocution of members of the public [29–31].

Miniature circuit breakers (MCBs) in a distribution board at the generator limit the electrical current drawn by each customer. The current ratings of the MCBs are used to determine the monthly charge and also provide overcurrent protection. Most customers buy less than 10 amperes to supply only their essential loads. **Figure 6** shows an MCB distribution board supplying around 100 dwellings installed at a neighbourhood diesel generator in Baghdad.

Figure 7 shows informal distribution circuits supplying power from a neighbourhood diesel in Baghdad. In IEK, it is common for individual final circuits to radiate directly from the generator distribution board to each consumer. Problems of the distribution circuits include loose or disconnected wiring, short circuits, mal-operating changeover switches and sustained high voltages caused by poor neutral earthing.

5. Operation of neighbourhood diesels

Neither the generators nor private wire distribution networks are regulated by the Federal Ministry of Electricity and the agreements between the generator operators and customers are verbal [32]. Customers sometimes experience poor power quality with voltage and frequency falling below their rated values of 230 volts and 50 Hz when the operators reduce the running speed of their engines to save fuel. It is also known for operators to overcharge their customers [5]. The disposal of engine lubricants in public sewage systems has been reported [33] while poor handling procedures of fuel and non-compliance with electrical safety regulations have been identified as causes of fires [23, 34, 35].

5.1 Noise

It is widely recognised that neighbourhood diesel generators can create a significant noise nuisance [8, 23, 36] especially when their enclosures and canopies are removed to increase cooling (**Figure 8**).

According to the Iraq 'Law of Noise Control' of 2015 [37], the noise limits in residential areas are Sound Pressure Levels (SPLs) between 55 and 60 dBA during the day and 45–50 dBA at night, depending on the source of the noise (e.g. local crafts and industrial workshops). There are no specific limits for the noise produced by neighbourhood diesel generators. In November 2019, the LPC in Erbil (Kurdistan Region of Iraq) issued new regulations requiring that, from May 2020, all neighbourhood diesels should be fitted with soundproofing systems. Non-compliant owners of the neighbourhood diesels face fines of approximately US $ 1670 and their licences being suspended [23, 38].

Figure 8.
Neighbourhood diesel generator with enclosures removed showing exposed fan (Source: Author).

The results of national studies of noise from neighbourhood diesels are shown in **Table 7**.

In [39], the SPLs of diesel generators with and without enclosures were measured at 1.1–1.2 metre from the ground. In [40] and [41], the SPLs of neighbourhood diesels installed in the cities of Duhok and Erbil were measured at various distances from the diesel generator. In [42], the SPLs produced by 250 kVA neighbourhood diesels were measured to investigate the impacts of noise pollution in the city of Mosul using geographic information systems (GIS).

Using the data from the references in **Table 7** and a simple hemi-spherical propagation model [43], Sound Power Levels for the generating sets were estimated of between 103 and 121 dBA without enclosures and about 91 dBA with an enclosure.

5.2 Emissions of CO, SO_2, NO_x, H_2S and total suspended particles (TSP)

The location of the neighbourhood diesel generators in residential areas leads to particular concerns over local air pollution. A 'Draft Iraqi Standard' [44] defines limits on the exhaust emissions from diesel generators but the common practice of mixing the diesel or gas oil fuel with heavy oil, and poor maintenance of the engines increase the level of emissions [23, 45, 46].

Table 8 shows emissions measured in local studies. The allowable emissions from small diesel generators are shown on the top line of **Table 8** with measured

Reference	Minimum SPL (dBA) at distance (m)	Maximum SPL (dBA) at distance (m)	Notes
[39]	63.1 dBA at 10 m with enclosure	89.2 dBA at 10 m without enclosure	Size of the generator sets not provided
[40]	74.86 dBA at 50 m	98.91 dBA at 5 m	State of the enclosure not available
[41]	69 dBA at 15 m	103 dBA at 1 m	Neither the number, rating nor the state of the enclosure available
[42]	63–65 dBA at 50 m	105–109 dBA at the generator site	The number of 250 kVA generator sets is not available. All units are without enclosures

Table 7.
Measurements of noise from neighbourhood diesels.

Reference			CO	NO_x	SO_2	H_2S	TSP
[44, 47, 48]	*Maximum permitted hourly concentrations of pollutants emitted from diesel generators according to the Draft Iraqi Standard*				*(ppm)*		
			0.26	**0.05**	**0.14**	**0.005**	—
[39]	*Month of 2012*	*August*	4.25	5.98	3.40	Not measured	0.48
		September	3.50	4.60	3.15		0.42
		October	2.95	3.90	2.44		0.32
		November	2.66	2.85	2.75		0.23
[47]	*Diesel generator rating (kVA)*	*150*	1.62–2.23	0.70	0.80–1.30	0.60–1.10	Not measured
		250	2.10–2.60	0.60–0.80	0.90–1.30	0.50–1.00	
		350	2.10	0.50	1.20	0.60	
		440	3.00	0.70	1.50	1.00	
		500	2.90–3.40	0.80–0.90	1.80–2.40	1.60–2.50	
		1000	3.10	0.90	2.10	2.10	
[48]	*College of Physical Education / Al-Qadisiya University*		2.80	Not measured	0.65	0.009	Not measured

Table 8.
Concentrations of pollutants emitted from different neighbourhood diesels in ppm.

values shown on the lower 3 lines. Reference [39] records the concentrations of air pollutants from diesel generators measured between August and November 2012. Higher wind speeds in autumn spread the pollutants and reduces their concentration. Alrawi and Hazim [47] show the maximum concentrations of CO, SO_2 and H_2S pollutants emitted from new and old 150, 250 and 500 kVA generators located in Baghdad. Najib [48] measured the emissions from diesel generators installed at Al-Qadisiya University. In all cases the measured emissions exceeded those specified in the draft standard.

6. Neighbourhood diesels in Kurdistan and other countries

In the Kurdistan Region of Iraq (KRI) in June 2020 consumers received an average of 16 hours per day of electricity from the public grid [49]. Neighbourhood diesels, however, remain common with at least 5500 generators registered and operating in the region [50–52]. Connection practice differs from elsewhere in Iraq with local distribution boards mounted on utility distribution poles from which the final connections radiate to customer premises. The boards are supplied using single conductor mains of 50–95 mm^2 copper conductor. The neutral wire connection practice (employing the neutral wire of the public grid) is similar to the practice seen in other Iraqi cities. The tariffs of the neighbourhood diesels in KRI are defined in ($/Amp) for neighbourhoods that have a connection to the public distribution grid. However, the tariffs are defined in ($/kWh) for the diesel generators supplying newly built residential housing complexes which are not connected to the public distribution grid [53, 54].

In Lebanon, neighbourhood diesels (known as ishtirak or 'subscription' [55]) have been common since the early days of the civil war of 1975–1990. In 2018, these generators, described by the World Bank Group as 'illegal and informal', were used to supplement customers with 8.1 TWh of power amounting to about 37% of the total power demand in Lebanon [56]. In Beirut, which has a daily supply of about 21 hours of electricity from the public distribution grid, neighbourhood diesels make up the 24-hour supply. In other cities of Lebanon which receive less than 12 hours of public grid electricity each day, the neighbourhood diesels supply customers with electricity for up to 6–8 hours per day [57–60].

The World Bank Group and the American University of Beirut [61] report that ratings of neighbourhood diesels in Lebanon are typically below 500 kVA, similar to Iraq. The connection practice of the neighbourhood diesels employing the neutral wire of the public distribution grid is the same [58]. Also, the contracts between the private entrepreneurs and the customers are verbal. Connection practice of the neighbourhood diesels in Lebanon is to use fuse boxes (or local distribution boards) mounted on subscribing buildings rather than on poles of the public distribution grid [57, 58]. Prior to October 2018 some customers only had MCBs while others had both MCBs and energy meters. Nowadays, all Lebanese customers (old and new) are required to have MCBs (to limit the maximum current) and energy meters (for tariff charging). There is also a standing charge defined by the current rating of a customer's MCB [62].

In Syria, neighbourhood diesels (locally called 'ampere or subscription' generators [63]) supply customers with electricity due to the damage sustained by the public grid during the civil war [64]. These generators were initially employed in regions controlled by the Syrian rebels to supply customers with no more than 10 hours of electricity per day [63]. The practice was later adopted in regions controlled by the Syrian Government [65, 66]. The topology of the private wire networks of the neighbourhood diesels in Syria is similar to KRI and Lebanon with thick single live conductors supplying local distribution boards mounted on public distribution poles or subscribing buildings. The use of the public network neutral wires is similar in Iraq, KRI and Lebanon [67]. The customers in Syria are not equipped with energy meters. The tariffs of the neighbourhood diesels, regulated by the LPCs in Syrian cities, are defined in ($/Amp).

7. Current status of rooftop solar PV systems in Iraq

Iraq, located between latitude 29°.98′ and 37°.15′, has a high potential of solar energy with a mean global PV potential of approximately 4.7 kWh/kWp, global horizontal irradiation (GHI) of 5.5 kWh/m^2 and an average of 3250 of hours of sunshine per year in Baghdad [68, 69] (**Figures 9** and **10**).

However, the utilisation of solar energy for electric power generation did not receive attention until 2019 when the Iraqi government (with the aid of international organisations) became more active in formulating a solar policy for the country [12]. Licences have been awarded for private companies to install residential solar power systems [71], technical specifications for these solar systems have been defined [72], and investors (local, international and IPPs) have been invited to construct grid scale solar plants [13] and pilot rooftop residential solar systems [73].

7.1 Specifications of rooftop solar systems

The technical specifications of rooftop solar PV systems issued by the Federal Ministry of Electricity imply that when the systems are financed by soft loans, they must be hybrid systems. Hybrid solar systems (**Figure 11**) combine the functions of solar panels, inverter, maximum power point tracker (MPPT), battery charger and

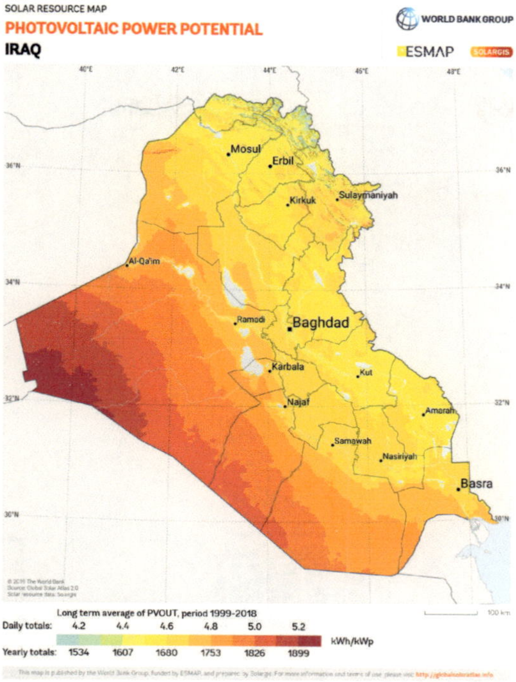

Figure 9.
PV power potential (PVOUT) in Iraq [70].

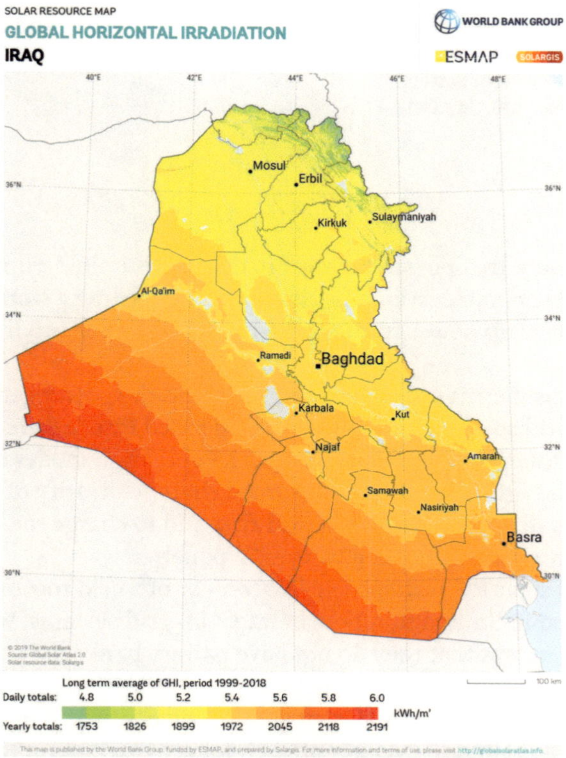

Figure 10.
Global horizontal irradiation (GHI) in Iraq [70].

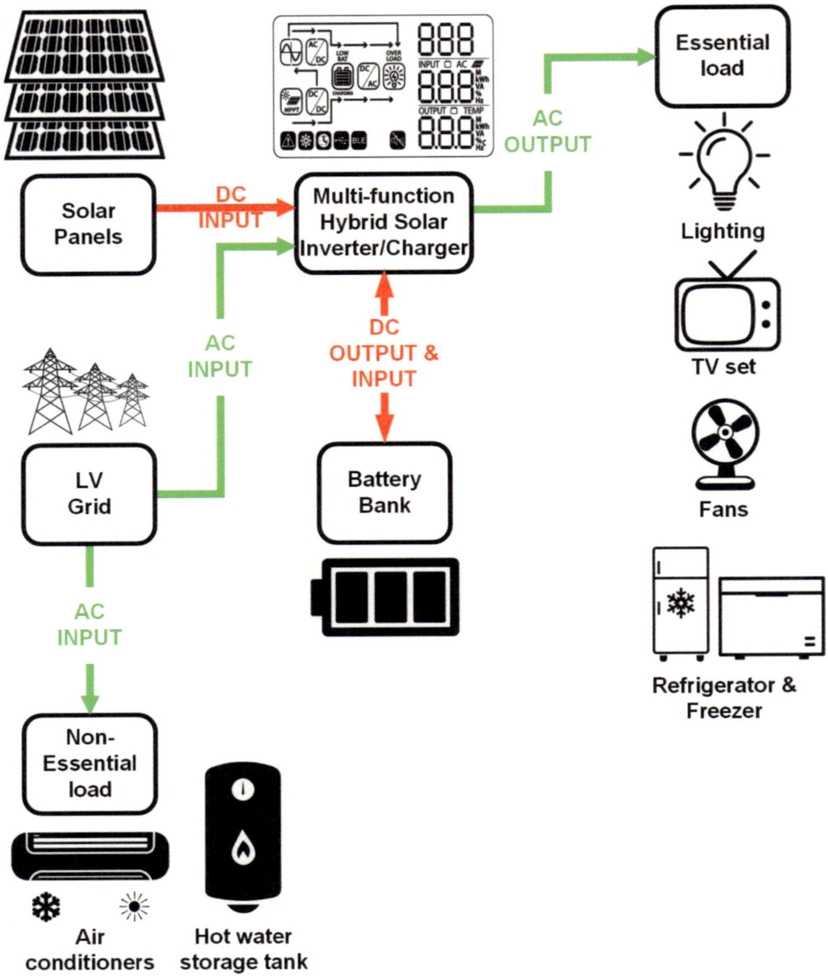

Figure 11.
Hybrid solar PV system.

battery pack to ensure that power supplied to the load is uninterrupted. A hybrid solar system can be operated as an on-grid system with battery storage or as an off-grid system with backup power from the grid. Power is never exported to the grid deliberately.

Taking into consideration the nature of loads and the power generation capacity (1–10 kW) of hybrid solar PV systems (recommended by the Federal Ministry of Electricity and commonly deployed in Iraq), the operation modes of these systems are summarised in **Figures 12–17**. It is assumed that the priority of a hybrid solar inverter/charger is to feed the essential load first and to charge the battery bank only if sufficient power is generated by the PV panels.

Besides hybrid solar PV systems, entirely on– or off–grid rooftop solar systems have been deployed in limited numbers in Iraq. On–grid systems, which are similar to hybrid systems except that they do not have battery banks, have been installed at a number of governmental buildings including the Federal Ministry of Electricity (an aggregate of 350 kW at two different sites), University of Babylon (with a 130 kW capacity) [74] and University of Technology [75].

In contrast, off–grid systems include battery banks, but are not connected to the LV distribution grid. Off–grid systems are used for rural agricultural (irrigation

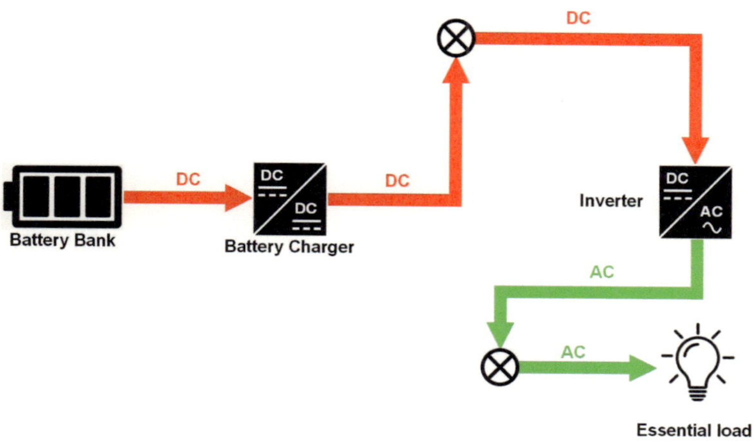

Figure 12.
Off-grid mode: PV power is not available. The essential load is fully supplied by the batteries.

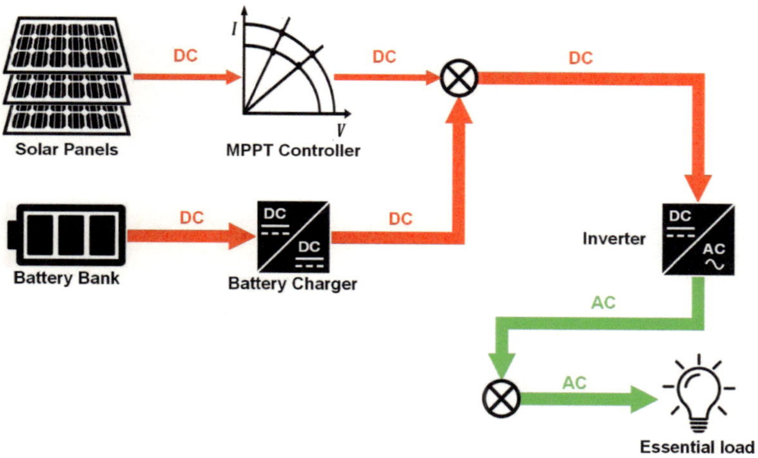

Figure 13.
Off-grid mode: PV power is not sufficient to supply the essential load which will therefore be supplied by both PV panels and batteries.

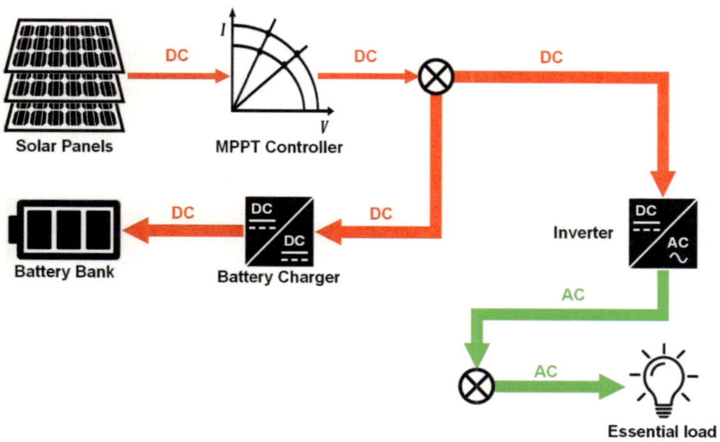

Figure 14.
Off-grid mode: PV power is sufficient to supply the essential load and charge the batteries.

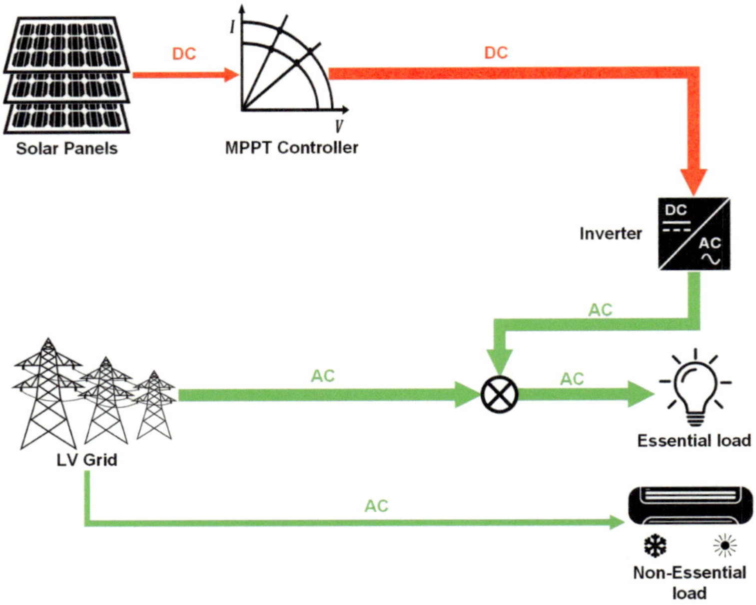

Figure 15.
On-grid mode: PV power is not sufficient to fully supply the essential load and the batteries are not connected (e.g. removed for maintenance or replacement). The essential load is supplied by both PV panels and LV grid.

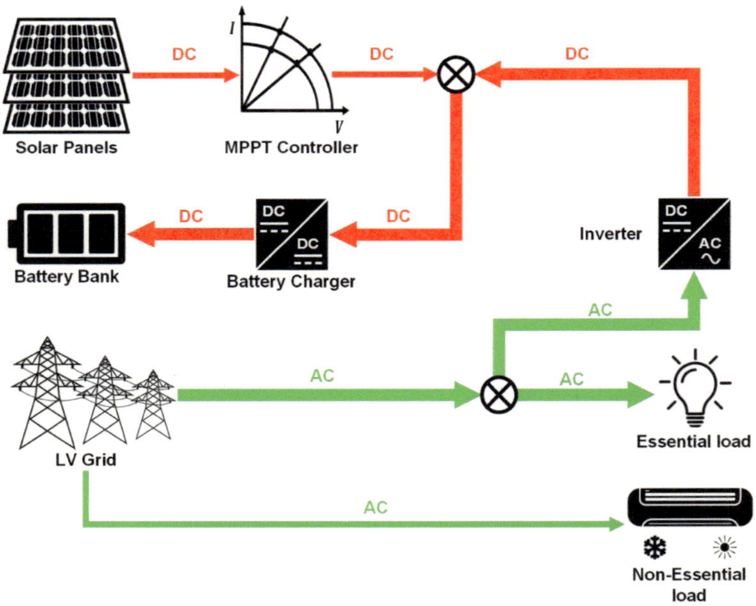

Figure 16.
On-grid mode: PV power is neither sufficient to supply essential load nor charge the batteries. LV grid supplies power to the essential load and charges the battery bank.

and drainage) applications and have also been employed for experimental studies. **Figure 18** shows an experimental off-grid system rooftop solar system installed at a residential premise in Baghdad.

A detailed illustration of the system is shown in **Figure 19**. Block (1) is the infeed cable collecting the outputs of the solar panels shown in **Figure 18**. Block (2)

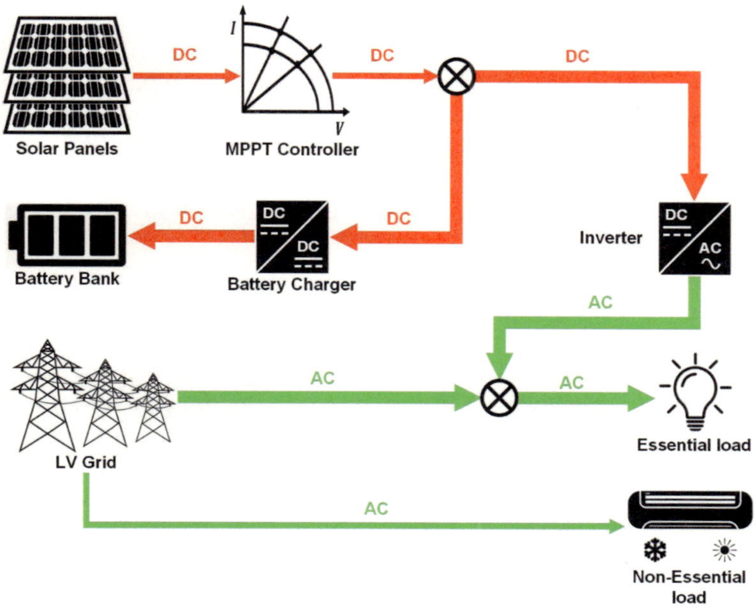

Figure 17.
On-grid mode: PV power is not sufficient to fully supply the essential load but is sufficient to charge the batteries. The essential load is supplied by both PV panels and LV grid.

Figure 18.
A rooftop array of solar panels in Baghdad (Source: Dr. Jaafar Ali Kadhum Al-Anbari).

Figure 19.
Detailed illustration of a 10 kW experimental rooftop off-grid solar system in Baghdad (Source: Dr. Jaafar Ali Kadhum Al-Anbari).

is a 10 kW inverter that converts 48 volts DC to 220 volts AC to supply the essential load of the residential premise. Block (3) is an MPPT charge controller while block (4) shows the cooling system installed to cool the inverter (block (2)). Finally, block (5) is a 48 volts battery bank comprising 54 lead acid batteries (of different capacities) connected to produce an aggregated capacity of 1500 Ah.

7.2 Rooftop solar panel systems as a sustainable source of power for Iraqi residences

Solar energy, if actively exploited, has an important role in improving Iraq's energy security and could help fill the gap between the available electrical power supply and demand without using traditional power generation technologies or neighbourhood diesel generators. Oil and gas consumed for power generation can be saved which in turn allows more oil exports that will add to the government revenues [12].

A pilot project comprising six rooftop solar PV systems (each having a capacity of 5 kW) in Najaf [76] was able, over four years, to save a total of 58 tonnes of CO_2 (equivalent of consuming more than 7000 gallons of diesel) from being emitted into the atmosphere [77]. Reference [78] reports that the potential savings in CO_2 emissions would amount to approximately 804 gCO_2/kWh should a 315 kW solar power plant be constructed at Sulaymaniyah airport to replace fossil fuel based electric energy supplying the airport.

A comparison between the present levelized cost of electricity (LCOE) from open-cycle gas turbines (OCGT), combined cycle gas turbine (CCGT), neighbourhood diesel generators and solar panels (**Table 9**) shows that rooftop solar PV systems offer a competitive alternative to neighbourhood diesel generators. In **Table 9**, residential rooftop solar systems have a maximum power generation capacity of 15 kW, commercial rooftop systems can generate up to 500 kW whereas utility-scale systems are multi-megawatt solar farms [79].

In Iraq, the installation cost of rooftop solar systems can either be paid as a one-off payment or over 36–60 months with a long-term loan. A 5 kW hybrid solar system costs between US $ 3800–4800 with a one-off payment whereas the cost of the same system increases to about US $ 6450 (over 36 instalments) – 6860 (over 60 instalments) on a long-term loan [80]. The variation in costs depends upon both the number of solar panels and batteries connected. A replacement lead acid battery is usually required every two years, at a cost of US $ 210–280 per 200 Ah battery.

Comparing the installation and battery replacement costs with the approximate electricity bill of a residential customer (**Table 6**), it can be concluded that a hybrid rooftop solar system is expensive and may not deliver the financial savings anticipated over its lifetime of 20–25 years. Similar findings were reported in [51]

Power generation technology		LCOE (US $/kWh)	Reference
OCGT		0.04–0.06	[5, 12]
CCGT		0.07–0.11	
Neighbourhood diesel		0.64–1.30	
Solar PV	**Utility-scale**	0.018–0.085	[5, 11, 12, 79]
	Commercial rooftop	0.062–0.064	[79]
	Residential rooftop	0.063–0.265	

Table 9.
Comparison between LCOE of solar PV and fossil fuel based power generation technologies.

recommending the installation of rooftop off-grid solar systems only when an annual discount rate of below 9.4% was assumed for the battery bank. Analysis of different scenarios showed that investment in rooftop solar systems would not be cost effective at high battery discount rates. Alternatively, reference [12] recommends exploring community solar microgrids rather than installations on each house.

In summary, it can be seen that numbers of rooftop solar system installations in Iraq are increasing; however, these will probably not reach a tipping point to replace neighbourhood diesel generators for some time. The public are often reluctant to install rooftop solar systems because of their high upfront and maintenance costs especially with the current unstable economic conditions in the aftermath of the coronavirus outbreak and worldwide drop in oil prices. The lack of government support for soft loan mechanisms as well as high commercial interest rates (more than 40%) for loans to fund domestic solar systems are other factors that discourage widespread installations of solar systems. Also, the customers are reluctant to invest in solar PV systems because present Iraqi legislations do not support net-metering or feed-in tariffs [12].

There is some evidence that the reducing cost of photovoltaic panels may offer a partial solution to this problem of deficit of generation. Iraq has an extremely attractive solar resource but so far implementation of photovoltaic generation has been limited. For widespread adoption of rooftop systems, a more attractive commercial climate is required, through low interest loans, net metering or feed-in tariffs.

8. Conclusions

The electricity systems of Iraq, and parts of Lebanon and Syria, experience frequent power cuts caused by shortage of generation, damaged transmission and distribution networks as well as rapidly increasing demand. In response to the limited hours that electricity is available from the public supply systems, local organisations have established innovative arrangements using diesel generators and simple distribution networks. These systems operate independently and are managed separately from the public electricity supply.

The generators are typically in the range of 100–500 kVA and are often locally manufactured from reused truck engines and imported generators. The generators provide each subscribing consumer with a supplementary supply of up to several kW of electrical power through informal networks that extend over a small area of a town or city. The final connection to the consumer premises is made through a radial single wire and the neutral of the public LV network. There is no connection of the live conductors from the generators with the public network and each customer has a changeover switch to select either the public mains when supply is available or the neighbourhood diesel. Monthly tariffs are based on $/amp with miniature circuit breakers limiting the current drawn by each consumer.

Neighbourhood diesels create significant local air pollution and noise, and can only supply small amounts of power at considerable cost. However, for those areas that have only limited public electricity supply they provide some power when the public service is unavailable. In Iraq, electricity from the public network is sold to domestic customers at a price that is below the cost of supply so limiting revenue that could be used to increase the capacity of the public supply system. There is no immediate prospect of the public electricity supply in Iraq improving dramatically and of these neighbourhood generators becoming redundant. Until the public electricity supply system can fully meet the load demand, the use of neighbourhood diesels is likely to continue.

Suitable Iraqi standards exist, some in draft form, to regulate the noise and gaseous emissions from neighbourhood diesels but local studies indicate these standards are not being met. No standards to regulate the novel connection practice of using a common neutral connection from the public network were identified. There appears to be scope both to enforce existing standards and develop a new electrical standard to regulate the connection and operation of the diesel generators and the innovative networks.

Acknowledgements

The author gratefully acknowledges the support of the FLEXIS project in the School of Engineering, Cardiff University. FLEXIS is part-funded by the European Regional Development Fund (ERDF), through the Welsh Government. Ariennir yn rhannol gan Gronfa Datblygu Rhanbarthol Ewrop drwy Lywodraeth Cymru. The author also acknowledges the assistance of Dr. Jaafar Ali Kadhum Al-Anbari providing the details and photos of an experimental 10 kW off-grid residential solar power system.

Author details

Ali Al-Wakeel
Cardiff University, Cardiff, United Kingdom

*Address all correspondence to: al-wakeelas@cardiff.ac.uk

References

[1] State Organization for Marketing of Oil (SOMO), "Iraq Crude Oil Exports - October 2020 (in arabic)," 2020. https://somooil.gov.iq/exports (accessed Nov. 27, 2020).

[2] Organisation of the Petroleum Exporting Countries (OPEC), "Iraq facts and figures," 2020. https://www.opec.org/opec_web/en/about_us/164.htm (accessed Sep. 15, 2020).

[3] Organisation of the Petroleum Exporting Countries (OPEC), "Oil Production Reduction Agreement," 2020. https://www.opec.org/opec_web/en/press_room/5891.htm (accessed Sep. 17, 2020).

[4] Central Intelligence Agency (CIA), "The World Factbook: Country Comparison: Natural Gas - Proved Reserves," 2018. https://www.cia.gov/library/publications/the-world-factbook/fields/273rank.html (accessed Sep. 15, 2020).

[5] International Energy Agency, "Iraq's Energy Sector: A Roadmap to a Brighter Future," 2019. [Online]. Available: www.iea.org/t&c/.

[6] United Nations Development Programme (UNDP) and Inter-Agency Information and Analysis Unit, "Electricity in Iraq Factsheet," 2010. [Online]. Available: https://reliefweb.int/sites/reliefweb.int/files/resources/F409BC15DE5570AC8525777C006B1430-Full_Report.pdf.

[7] United Nations Assistance Mission for Iraq and United Nations Development Programme (UNDP), "Overview of Iraq's Electricity," 2008. [Online]. Available: https://iraqslogger.powweb.com/downloads/Overview_of_Iraq_Electricity.pdf.

[8] Iraq Prime Minister Advisory Commission, "Integrated National Energy Strategy (INES): Final Report," 2012. [Online]. Available: http://documents1.worldbank.org/curated/en/406941467995791680/pdf/105893-WP-PUBLIC-INES-Summary-Final-Report-VF.pdf.

[9] Attaakhi Newspaper, "Private and government owned generators as many as Iraq's palm trees imposing their powers," *7030*, p. S-1, Dec. 20, 2015.

[10] P. Fairley, "Rooftop Solar Takes Hold in Iraq in the Aftermath of ISIS," *IEEE Spectr.*, no. June, pp. 9-10, 2018.

[11] H. Istepanian, "Solar Energy in Iraq: From Outset to Offset," 2018. [Online]. Available: https://iraqenergy.org/product/solar-energy-in-iraq-from-outset-to-offset-report/.

[12] H. Istepanian, "Iraq Solar Energy: From Dawn to Dusk," 2020. Accessed: Oct. 03, 2020. [Online]. Available: http://library.fes.de/pdf-files/bueros/amman/16324-20200722.pdf.

[13] Iraqi Ministry of Electricity, "Call for Expressions of Interest for Solar Energy Independent Power Producer (IPP) Projects," 2019. http://www.mofa.gov.iq/zagreb/wp-content/uploads/sites/30/2019/05/536.pdf (accessed Sep. 15, 2020).

[14] Yesar Al-Maleki, "Overview of Iraq's Renewable Energy Progress in 2019," 2020. https://iraqenergy.org/2020/02/20/overview-of-iraqs-renewable-energy-progress-in-2019/ (accessed Oct. 03, 2020).

[15] Shafaq News Agency, "The Ministry of Electricity of Kurdistan Region: We make use of half of production capacity of 7000 MW," 2019. https://shafaaq.com/en/kurdistan/the-ministry-of-electricity-of-the-region-we-make-use-of-half-of-production-capacity-energy-of-7000-mw/ (accessed Sep. 15, 2020).

[16] Iraqi Ministry of Electricity, "2018 Statistical Report - Generation, Transmission and Distribution (in arabic)," 2018. [Online]. Available: https://www.moelc.gov.iq/home/part/annual-reports/view/home-news-1579500207.

[17] Iraqi Ministry of Electricity and Parsons Brinckerhoff, "Iraq Electricity Masterplan," 2010. Accessed: Sep. 17, 2020. [Online]. Available: http://www.iraq-jccme.jp/pdf/archives/electricity-master-plan.pdf.

[18] Inter-agency Information and Analysis Unit (IAU) *et al.*, "Iraq Knowledge Network: Essential Services Factsheet," 2011. [Online]. Available: https://reliefweb.int/sites/reliefweb.int/files/resources/ServicesFactsheet-English.pdf.

[19] Al-Araby Newspaper, "Iraq: A hard summer accompanied with power cuts and lockdown measures (in arabic)," Jul. 08, 2020.

[20] Xinhua News Agency, "Shadows of electricity crisis casting again in Iraq as temperatures soar (in arabic)," Jul. 14, 2020.

[21] Iraqi News Agency (INA), "Baghdad Governor: Mechanism of distributing free fuel rations amongst private neighbourhood generators (in arabic)," Jul. 14, 2020. https://www.ina.iq/109730/ (accessed Sep. 20, 2020).

[22] Directorate of Electricity Commission, *Regulations of Electric Power Supply (in arabic)*. Iraq: Iraqi Commission of Electricity, 1999, pp. 1-26.

[23] Rudaw News Agency, "Deadly generators patch up Kurdistan's power supply." https://www.rudaw.net/english/kurdistan/11082020 (accessed Sep. 15, 2020).

[24] Baghdad Provincial Council, Kirkuk Provincial Council, Diyala Provincial Council, Karbala Provincial Council, and Dhiqar Provincial Council, *Regulatory instructions of neighbourhood diesel generators (official letters in arabic)*. 2017.

[25] The World Bank Group, "Iraq economic monitor: from war to reconstruction and economic recovery," 2018. [Online]. Available: http://documents.worldbank.org/curated/en/771451524124058858/pdf/125406-WP-PUBLIC-P163016-Iraq-Economic-Monitor-text-Spring-2018-4-18-18web.pdf.

[26] Iraqi Ministry of Electricity, "Electricity tariffs (in arabic)," Sep. 19, 2020. https://www.moelc.gov.iq/home/page/electricity_tariffs_fees?lang=ar (accessed Sep. 19, 2020).

[27] Oil Products Distribution Company (OPDC), "Neighbourhood Generators' Rations of Gas Oil (in arabic)," Nov. 07, 2019. https://opdc.oil.gov.iq/index.php?name=News&file=article&sid=756 (accessed Sep. 19, 2020).

[28] Iraqi Ministry of Electricity, "Estimating consumption and calculating bills of electrical energy (in arabic)," 2020. https://www.moelc.gov.iq/calculator?lang=ar (accessed Sep. 19, 2020).

[29] China Central Television (CCTV), "The phenomenon of random electricity wires in Iraqi neighbourhoods (in arabic)," Aug. 26, 2016. http://arabic.cctv.com/2016/08/26/VIDEo2jxLNoeqyofpxVbfGWF160826.shtml (accessed Sep. 20, 2020).

[30] Attaakhi Newspaper, "Private and government neighbourhood generators: A chronic disease speeding up the death of Iraqi citizens (in arabic)," *7030*, p. S-4, Dec. 20, 2015.

[31] Radio Free Iraq (Radio Free Europe/Radio Liberty), "Legislative list to regulate the operation of private neighbourhood generators in Baghdad

(in arabic)," 2013. https://www.iraqhurr.org/a/24924146.html (accessed Sep. 20, 2020).

[32] Al-Sumaria News Agency, "The electricity and the art of illegal use of fuel .. 'mafias' whose victims are helpless citizens: Statement by MP Jamal Kochar (in arabic)," 2016. https://www.alsumaria.tv/mobile/news/183627/ (accessed Sep. 21, 2020).

[33] Mayoralty of Baghdad, "Mayoralty of Baghdad urges private entrepreneurs to comply with environmental regulations (in arabic)," 2017. https://amanatbaghdad.gov.iq/posts.php?lang=ar&post=401 (accessed Sep. 21, 2020).

[34] Iraqi Parliament, "Fires in Iraq (in arabic)," 2013. [Online]. Available: http://parliament.iq/wp-content/uploads/2017/05/

[35] A.-Q. A.-R. Fadhil, "Fire losses in Iraq (in arabic)," 2014. https://iraqinsurance.wordpress.com/2014/02/13/fire-losses-in-iraq/ (accessed Sep. 21, 2020).

[36] Iraqi Ministry of Environment, United Nations Development Programme (UNDP), United Nations Environment Programme (UNEP), and World Health Organisation (WHO), "The National Environmental Strategy and Action Plan for Iraq (2013-2017)," 2013. Accessed: Sep. 21, 2020. [Online]. Available: http://wedocs.unep.org/handle/20.500.11822/8726.

[37] Presidency of Iraqi Government, *Law of Noise Pollution Control (in arabic)*. Iraq: Office of President of Iraq / Ministry of Justice, 2015, pp. 1-9.

[38] Rudaw News Agency, "New Erbil governor pledges crackdown on noisy, dirty private generators," 2019. https://www.rudaw.net/english/lifestyle/24112019 (accessed Oct. 10, 2020).

[39] S. S. Al-Kizwini, A. K. Idrees, and R. S. Mehdi, "Study of the environmental pollution impacts of electrical generators upon their surroundings diesel and gasoline generators (in arabic)," *J. Univ. Babylon*, vol. 21, no. 5, pp. 1705-1721, 2013, Accessed: Sep. 21, 2020. [Online]. Available: https://www.iasj.net/iasj/download/429df05474216e3b.

[40] M. A. Mahammed, F. A. Kochery, and M. S. Abdulkhaliq, "Investigation of noise pollution of electrical diesel generators in Duhok city/ Kurdistan of Iraq," *J. Univ. Zakho*, vol. 1, no. 1, pp. 319-324, 2013.

[41] S. Saber, "Environmental noise with solutions: A case study," *Int. J. Adv. Appl. Sci.*, vol. 1, no. 2, pp. 6-14, 2014, Accessed: Sep. 21, 2020. [Online]. Available: http://www.science-gate.com/IJAAS/Articles/2014-1-2/04 2014-1-2-pp.6-14.pdf.

[42] S. S. Q. Al-Tae'e and A. Z. A. Salih, "Study of the environmental pollution of private generators in the right-side of Mosul using GIS (in arabic)," *J. Tikrit Univ. Humanit.*, vol. 19, no. 8, pp. 360-385, 2012.

[43] W. T. W. Cory, "Relationship between Sound Pressure and Sound Power Levels," in *Eurovent WG 1*, 2006, pp. 4-9, [Online]. Available: http://www.eurovent-certification.com/fic_bdd/pdf_fr_fichier/1137149375_Review_67-Bill_Cory.pdf.

[44] Central Organisation for Standardisation and Quality Control (COSQC) - Iraqi Ministry of Planning, "Minutes of Standards Accreditation Panel Meeting (in arabic)," 2013. http://cosqc.gov.iq/ar/newsdetails.aspx?NID=124 (accessed Sep. 30, 2018).

[45] United States Government Accountability Office, "Securing, Stabilizing, and Rebuilding Iraq: Key Issues for Congressional Oversight," 2007. [Online]. Available: https://www.gao.gov/assets/210/203060.pdf.

[46] World Bank, "Integrated National Energy Strategy (INES): final report (English)," 2012, [Online]. Available: http://documents.worldbank.org/curated/en/406941467995791680/Integrated-National-Energy-Strategy-INES-final-report.

[47] A. K. Alrawi and R. Hazim, "Studying the environmental impacts of electrical generators in Baghdad / Al-Karrada district 903 (in arabic)," *J. Coll. Educ. Wasit Univ.*, vol. 14, pp. 294-314, 2013.

[48] A. A. Najib, "The impact of exhaust gases emitted from electric generators on some physiological variables of female students of the second stage in the Faculty of Physical Education, Al-Qadisiya University (in arabic)," *Sci. Phys. Educ. - Babylon Univ.*, vol. 9, no. 1, pp. 40-54, 2016.

[49] KRG Ministry of Electricity, "250 MW added to the public grid in July 2020 (in Kurdish)," 2020. http://moel.gov.krd/Direje.aspx?Jimare=2986 (accessed Sep. 21, 2020).

[50] Kurdistan 24 News Agency, "Kurdistan officials warn of alarming levels of air pollution in Duhok," 2019. https://www.kurdistan24.net/en/news/34c679d8-bf5e-4dd9-adaf-06bd755b6094 (accessed Sep. 21, 2020).

[51] D. Baban and P. Askari, "Future Sustainable Energy Solutions for Sulaymaniyah," KTH ROYAL INSTITUTE OF TECHNOLOGY, 2019.

[52] European Union (EU), Kurdistan Regional Government (KRG), United Nations Development Programme (UNDP), and Covenant of Mayors for Climate & Energy, "Sustainable Energy Action Plan (SEAP) : Kurdistan Region of Iraq - Erbil Governorate," 2018. Accessed: Sep. 21, 2020. [Online]. Available: https://www.climamed.eu/wp-content/uploads/files/Iraq_Erbil-Governorate_SEAP.pdf.

[53] Zamen News Agency, "Disagreements between Ministry of Electricity and the Council of Kurdistan Provinces upon tariffs of residential housing complexes (in Kurdish)," Feb. 06, 2020. https://www.zamenpress.com/Print.aspx?jimare=16997 (accessed Sep. 23, 2020).

[54] KNN News Agency, "Tariff of private neighbourhood generators for May 2020 (in Kurdish)," Jun. 02, 2020. https://www.knnc.net/Details.aspx?jimare=30064 (accessed Sep. 23, 2020).

[55] Ramboll Group and U.S. International Development Finance Corporation, "ESIA REPORT: LEBANON WIND POWER WIND FARM, LEBANON," May 2019. Accessed: Sep. 23, 2020. [Online]. Available: https://www3.opic.gov/Environment/EIA/lebanonwindpower/ESIA.pdf.

[56] A. Ahmad, World Bank Group (Energy & Extractives), and Energy Sector Management Assistance Program (ESMAP), "DISTRIBUTED POWER GENERATION FOR LEBANON: Market Assessment and Policy Pathways," May 2020. Accessed: Sep. 23, 2020. [Online]. Available: https://openknowledge.worldbank.org/bitstream/handle/10986/33788/Distributed-Power-Generation-for-Lebanon-Market-Assessment-and-Policy-Pathways.pdf?sequence=1&isAllowed=y.

[57] World Bank, "Lebanon Social Impact Analysis - Electricity and Water Sectors," *World bank Rep.*, vol. 48993-LB, no. 48993, pp. 1-83, 2009, [Online]. Available: http://documents.worldbank.org/curated/en/153931468263692245/Lebanon-Social-impact-analysis-electricity-and-water-sectors%0Ahttp://siteresources.worldbank.org/INTPSIA/Resources/490023-1120841262639/Lebanon_electricity_water.pdf.

[58] E. Verdeil, "Beirut: Metropolis of Darkness and the Politics of Urban

Electricity Grids," in *Energy, Power and Protest on the Urban Grid*, First Edit., Routledge Taylor and Francis Group, 2016, pp. 155-175.

[59] United Nations Refugee Agency (UNHCR), United Nations Children's Fund (unicef), World Food Programme (WFP), and Inter Agency Coordination (IAC) Lebanon, "Vulnerability Assessment of Syrian Refugees in Lebanon," 2019. Accessed: Sep. 23, 2020. [Online]. Available: http://ialebanon.unhcr.org/vasyr/#/.

[60] International Monetary Fund (IMF), "Lebanon : Selected Issues," Oct. 2019. Accessed: Sep. 23, 2020. [Online]. Available: https://www.imf.org/~/media/Files/Publications/CR/2019/1LBNEA2019002.ashx.

[61] A. Shihadeh *et al.*, "Effect of Distributed Electric Power Generation on Household Exposure to Airborne Carcinogens in Beirut," Beirut, 2013. [Online]. Available: https://scholarworks.aub.edu.lb/bitstream/handle/10938/21130/20130207ifi_rsr_cc_effect_Diesel.pdf?sequence=1&isAllowed=y.

[62] Lebanese Ministry of Energy and Water, "Private generator tariffs - October 2018 (in arabic)," 2018. http://www.energyandwater.gov.lb/.

[63] Assistance Coordination Unit (ACU), "Harim District Dynamo Report (March 2018)," pp. 1-73, 2018, [Online]. Available: https://reliefweb.int/sites/reliefweb.int/files/resources/Harim DYNAMO_En_040418.pdf.

[64] Youssef Diab, Abboud Hajjar, and Oliver Waine (translation), "Electricity in War-Torn Aleppo: A New Form of Urban Management (translated from French)," *Metropolitics*, Nov. 2018, Accessed: Sep. 26, 2020. [Online]. Available: https://metropolitics.org/Electricity-in-War-Torn-Aleppo-A-New-Form-of-Urban-Management.html.

[65] World Bank Group, "SYRIA DAMAGE ASSESSMENT of selected cities Aleppo, Hama, Idlib," 2017. Accessed: Sep. 26, 2020. [Online]. Available: http://documents1.worldbank.org/curated/en/530541512657033401/pdf/121943-WP-P161647-PUBLIC-Syria-Damage-Assessment.pdf.

[66] F. Witi, "Ampere generators in Idlib countryside (in arabic)," 2016. https://geiroon.net/archives/63038 (accessed Sep. 27, 2020).

[67] Al-Souria News Agency, "The power cuts in Aleppo make generators an indispensable necessity (in arabic)," Feb. 28, 2015. https://www.alsouria.net/archive/content/ (accessed Sep. 27, 2020).

[68] Weather & Climate, "Average monthly hours of sunshine in Bagdad, Iraq," 2020. https://weather-and-climate.com/average-monthly-hours-Sunshine,Bagdad,Iraq (accessed Oct. 03, 2020).

[69] The World Bank and Energy Sector Management Assistance Program (ESMAP), "GLOBAL PHOTOVOLTAIC POWER POTENTIAL BY COUNTRY," Jun. 2020. Accessed: Oct. 03, 2020. [Online]. Available: https://solargis.com.

[70] The World Bank Group, Global Solar Atlas, and Energy Sector Management Assistance Program (ESMAP), "Iraq - Solar Irradiation And PV Power Potential Maps," Aug. 22, 2018. https://globalsolaratlas.info/download/iraq (accessed Oct. 04, 2020).

[71] Iraqi Ministry of Electricity, "Instructions for installing solar energy systems for citizens, using soft loans, and the mechanism for qualifying the companies to implement these projects (in arabic)," 2019. https://www.moelc.gov.iq/attachments/info_1.pdf (accessed Oct. 04, 2020).

[72] Iraqi Ministry of Electricity, "Specifications of residential rooftop

solar power systems (in arabic)," 2019. https://www.moelc.gov.iq/attachments/969.pdf (accessed Oct. 04, 2020).

[73] Iraqi Ministry of Electricity, "Supplying, execution and installing operating solar energy systems for citizens in six governorates (in arabic)," 2020. https://www.moelc.gov.iq/home/tender/view/home-tender-1583915929?lang=ar (accessed Oct. 04, 2020).

[74] Iraqi Ministry of Electricity, "Renewable energy services (in arabic)," 2019. https://www.moelc.gov.iq/home/page/sustainable_energy?lang=ar (accessed Oct. 07, 2020).

[75] Energy and Renewable Energies Technology Center (ERECT) / University of Technology, "Guide of Energy and Renewable Energies Technology Center (in arabic)," 2017. https://eretc.uotechnology.edu.iq/images/sj_univer/pdf/dalel2016.pdf (accessed Oct. 07, 2020).

[76] United Nations Development Programme (UNDP), "Catalysing the Use of Solar Photovoltaic Energy," 2015.

[77] United Nations Development Programme (UNDP), "Testing the potential of solar in Iraq," 2020. https://www.iq.undp.org/content/iraq/en/home/stories/2020/06/testing-the-potential-of-solar-in-iraq.html (accessed Oct. 13, 2020).

[78] K. Khan, A. Azabany, and W. Ahmed, "Reductions in CO2 emissions from electricity generation from solar energy at Sulaymanyah airport in Kurdistan, Iraq," *Asian J. Sci. Technol.*, vol. 5, no. 8, pp. 479-481, 2014, [Online]. Available: https://www.journalajst.com/sites/default/files/issues-pdf/1635.pdf.

[79] International Renewable Energy Agency (IRENA), "Renewable Power Generation Costs in 2019," Abu Dhabi, 2020. Accessed: Oct. 08, 2020. [Online]. Available: https://www.irena.org/publications/2020/Jun/Renewable-Power-Costs-in-2019.

[80] International Islamic Bank, "Prices of solar energy systems (in arabic)," 2020. https://imtb.iq/services/solar/ (accessed Oct. 09, 2020).

Chapter 7

Regulatory Impediments to Micro-Wind Generation

Ryan M. Yonk, Corbin Clark and Jessica Rood

Abstract

Recent growth in the renewable energy industry has largely been driven by government support for alternative energy. Wind power in the United States is the second largest source of renewable energy, and has been heavily subsidized by state and federal government. There has also been an increasing interest in small scale environmental community projects, and this trend is expected to continue. Currently, there are 2 terawatt hours (TWh) of potential energy capacity through small- and micro-wind projects throughout the United States. Increased development of micro-wind energy could significantly impact America's non-hydropower renewable energy generation. Micro-wind, the utilization of the flow of wind energy to produce electricity for a house, farm or other non-utility scale generation can be regulated at the federal level, as well as at the state and local/community level. We examine two cases of micro-wind energy production to explore the regulatory impediments these smaller projects face. We find that the level of complexity of the regulatory framework is discouraging for innovation and development, and that the benefits of installing energy-generation are often outweighed by the cost of implementation.

Keywords: micro-wind, regulation, renewable energy, regulatory systems, innovation, technology

1. Introduction

Over at least the last twenty years, a substantial and concerted effort has been made to remake the generation of electrical power in the United States. The calls-to-action to move away from fossil fuel-based generation of electricity have commonly called for "green energy", "alternative energy" or "renewable energy" as the replacement. Those terms have faced both widespread adoption and rabid disagreement over which power generation sources and scope should be prioritized as the replacement of fossil fuels. While in practice, these terms are used interchangeably among both policy makers and the general public, for those who choose one term over another, they often represent nuanced differences on what should or should not be included among the alternatives under consideration. Nuclear power generation often falls within this distinction with some suggesting it as a carbon reducing alternative and others pointing to the environmental risks posed by nuclear power generation, such as the disposal of waste products from that generation. Likewise, hydropower, particularly large-scale projects, face similar concerns and complaints from some who advocate for a large-scale movement away from fossil fuels, and the subsequent replacement of fossil fuels with other alternatives. In this chapter we use the term renewable energy or renewables to represent the relatively wide swath of non-fossil fuel alternatives [1].

Previous work has explored these controversies in detail and we do not endeavor to recreate this discussion. We will maintain an agnostic position with regards to what term, or power sources ought to be considered, or the necessity of large-scale conversion from fossil fuels in this chapter. We instead explore regulatory burdens and impediments that are faced by the development of micro-wind generation approaches.

Our exploration reviews the regulatory process within larger efforts by the United States to increase renewable energy use. This combined regulatory analysis allows us to explore the impact of regulation and policies supporting renewable energy on the development of micro-wind systems in the United States. We first review the history of these production approaches and then focus on the effects regulation has on micro-wind generation. We use the framework developed in previous work on micro-hydropower and published in "The Regulatory Noose: Logan City's Adventures in Micro-Hydropower" [2].

2. Background

The literature on the impact of regulation has been well documented and thoroughly explored. Scholars have detailed the direct effect of federal regulation, particularly on the economy. Within the broader literature, a substantial critical evaluation has explored the efficacy of the political process in making policies that achieve their stated purpose and avoid unintended consequences. These reviews have generally found that regulation, including federal regulation, faces substantial problems in achieving these dual purposes. One of the most commonly identified problems comes from Tullock [3]. He identified "Rent-Seeking" and the resulting distortions to the policy outcome and decision-making process as one reason political processes are ill-equipped to effectively create policies. Rent-seeking identifies that in order to achieve "rents," some economic or political gain, interest groups are willing to use resources (including economic resources) to influence the regulatory process in their favor. They do so to ensure that long run policy and economic profits are protected. Policies including regulatory preference or expansion, market limitations through tariffs and other restrictions on trade can be obtained from political agents, and those with vested interests face strong incentives to engage. Further expanding in this area, Stigler put forth a strong theoretical frame that is rooted in similar thinking which suggests that the supply and demand for regulatory action is heavily influenced by special interest groups who want public resources and protection that can be supplied by public officials [4].

While rent-seeking models are useful in explaining the adoption and creation of preferential regulatory systems, those systems tend to persist in the face of alternative pressure from other sources. Reversing prior decisions, especially regulatory requirements, has been demonstrated to be particularly problematic. One examination of technological requirements within regulation is illustrated by Arthur [5]. He argues that after a technology has been adopted by its users, there are increasing returns to its use, leading to stronger preferences for continued use of that same technology. As the time horizon extends, the more experience with and adaptation of the regulatorily preferred or required technology increases. As a result, the particular technology becomes stuck; decreasing the probability that it is or will be interchanged with other technology. The cost of switching to a different (possibly more efficient, cost effective) technology becomes prohibitive and the regulatory requirement becomes the default preferred approach. This phenomenon is commonly referred to as path-dependence, and is one of the core mechanisms whereby rent-sought regulatory outcomes become difficult to change even when benefits to alternative arrangements are available.

Path dependence has been well documented within the policy sphere and institutional economics, particularly, historical institutionalism has clearly demonstrated both the propensity for path dependence to emerge, and the costs associated with that path dependence. Arrow's exploration of path dependence highlights the similarities of increasing returns in the costs of establishing and sustaining an institution that relies on technology [6, 7]. These costs are difficult to escape, especially when technology has been regulatorily determined. Pierson explains that political institutions are especially vulnerable to falling into path-dependence [8]. By their nature, bureaucracies tend to grow in both size and scope. Bureaucrats, who often are drawn to their agency due to interest and expertise, seek to influence their particular area and to do so, push for increased influence and regulation. Increased influence and regulatory involvement requires increased authority and larger budgets. This natural push leads to increasingly complex hierarchical webs and complicated regulatory standards. As McClaughlin and Williams point out, bureaucracies continue to pile new regulations on top of the old, further confounding regulations through which individuals and developers must wade through [9]. This stifling regulation hinders innovation and evolution because of increased cost in terms of dollars and time. The end result is a regulatory regime that is neither efficient or efficacious in achieving the goals laid out, but rather serves to protect both the rent-seeker and the rent-providers who are linked in mutually beneficial arrangements that ultimately stifle change and adaptation outside of the artificially created regulatory ecosystem.

In previous work by one of the authors of this chapter, Green vs. Green, he and his co-authors explore this reality by examining the landscape of the environmental regulatory web that green energy producers face [10]. After describing the development of environmental regulations, they provide both an approach to examining the effect of regulation on green energy projects and examples of cases where projects were disrupted. This chapter expands on that approach and applies that lens to micro-generation of wind power. We find that while legislation intends to bolster green energy production, it instead actively increases the costs of green micro-wind projects especially when coupled with other regulatory rules.

3. Introducing micro-wind

Wind energy is the second most utilized renewable energy source in the United States, with 338 billion kilowatt hours (kWh) in 2020, up from 6 billion kWh in 2000 [11]. Advances in technology used to produce wind energy have decreased the cost of producing electricity from wind. Along with the improvement in technology, government programs directed toward increasing green energy have contributed to make wind energy one of the fastest growing industries in the nation. These improvements in wind power are becoming more relevant as the literature finds that alternative energy sources are crucial to the future of both the environment and economy of the United States.

Micro-wind is the most accessible form of micro-energy, and the simplest form of clean energy. Wind turbines work by harnessing the kinetic energy that wind creates. A turbine has blades that function similarly to airplane wings; when wind flows over them, they create lift causing the blade to turn. The blades are connected to a drive shaft that spins an electric generator which produces electricity [12]. Micro-wind is similar to large-scale wind energy production, simply on a smaller scale. Instead of having farms of massive wind turbines, a micro-wind project could include one or more turbines connected to a relatively small generator. Micro-wind is suitable for residential energy production, used mostly for a house or farm. The excess energy can then be exported to the electrical grid, and credits can be provided by the retailer.

There are numerous potential benefits to using wind turbines and generating clean energy. Wind energy proponents claim that it produces no air, water, or thermal pollution, nor greenhouse gases, and no smog. Advocates claim that they leave few impacts on the local environment when they are dismantled, and that most activities on a wind site are not halted due to wind installations. These claims, however, are not with-out controversy and some evaluations have sound significant environmental impacts of wind generation [10]. The price of energy can be competitive, and it is quickly built and installed due to the straightforward design. The initial cost of installation of micro-wind turbines can be higher than that of other energy sources, but in the long-run they have the potential to be cost-effective [13].

Micro-wind is being looked at on a smaller scale than other micro-energy sources, with residential areas using micro-wind at the highest level. Small wind refers to small turbines that typically exert power of between 500 W and 25 kW, which may or may not be hooked up to the grid. Unlike hydro power, wind energy can be harvested from virtually everywhere, with manufacturers recommending a minimum average wind speed of 4.5 to 5 m/s [14]. One of the major benefits from micro-wind power is that it allows for the extension of clean, renewable energy to areas with limited to no grid access [15].

Use of micro-wind has increased dramatically over the past decades, with experts estimating that over one million micro-wind turbines are in use globally [15]. The complications of regulations in the micro-wind energy generation often scare away city officials and homeowners from implementing this technology on a wider scale. Without substantial regulation on micro-wind turbines, many more cities and consumers might choose to utilize the technology, potentially decreasing the environmental footprint, and saving the consumer money. Installation cost for micro-wind is relatively low, and maintenance is comparable to current costs, and is often less common. Wind turbines are not as efficient as other green options (the average wind efficiency of turbines falling between 35% and 45%), but with the low costs mentioned above, they may still be a viable option. At the speed of innovation in the past decades, some have claimed that the rate of efficiency will climb higher. If this claimed rise in efficiency occurs, micro-wind could potentially see greater demand become more widespread and help the United States increase renewable energy production with an efficient dispersed source. While the technical challenges may have solutions likely to occur in the near term, the regulatory environment faced by micro-wind remains daunting, and likely to prevent widespread adoption.

4. Current micro-wind regulation

Micro-wind, unlike the more common micro-hydro, is more isolated from the grid and focused on residential power creation, but that seeming isolation does not necessarily mean that regulation does not exist. Micro-wind regulation is primarily concerned with physical limitations and construction rather than the other more technical aspects which are more relevant to micro-hydro [2]. Many residential areas have zoning, permitting, and covenant guidelines that must be adhered to including, but not limited to, height limits, which proves to be a serious issue for accessing higher speed winds, and noise level caps. The noise issue is not as serious an impediment as height because the ambient noise of a micro-wind turbine is only slightly above that of natural ambient wind [16]. There are at least 250 state policies regulating the construction of only small wind projects, with nearly double that applying to both large and small projects.

The current process for permitting and installing a wind turbine can be long and arduous. While the initial steps of assessment can be costly, they are primarily separate from the regulatory system [17]. The following stage is dedicated to complying with federal regulations as well as local government zoning, permitting, and covenant requirements. As a result, the regulatory environment may differ on the federal, state, county, and potentially sub-county levels. Meaning that every sight faces layers of regulation that can be daunting to navigate. In addition, residential micro-wind homeowner's associations may further complicate producing wind energy.

As we note one of the primary complications for the use of micro-wind energy is the necessity of meeting not just local and state requirements but meeting federal regulations that were designed for large-scale projects, and which often did not consider the possibility of smaller projects. Despite some attempts to reduce regulatory burden of interconnections particularly IEEE's 1547 standard which is designated under the Energy Policy Act of 2005 concerns about the applicability of the section remain for micro-wind [18]. We explore the application of these regulations and identify others that might potentially impact micro-wind depending on site specific considerations.

The Federal Energy Regulatory Commission (FERC) is the government regulatory commission that oversees and is responsible for determining what level of analysis is required for a given energy project. Generally, FERC oversees the grid connections and generators connected to the higher voltage systems, while states regulate the retail markets for electricity and oversee the connectivity of generators connected to lower-voltage systems. The commission's oversight includes installing and ensuring the compliance with a laundry list of legislation that includes the Energy Policy Act of 2005, the Federal Deepwater Port Act, the Endangered Species Act, the Fish and Wildlife Coordination Act, the National Environmental Policy Act of 1969 (NEPA), the National Historic Preservation Act, the Rivers and Harbors Act, and the Wild and Scenic Rivers Act [19]. These policies contribute to the difficulty of implementing micro-wind power for consumers.

When applying for a license with FERC, even a small turbine must be registered and adhere to all the guidelines applicable to wind farms. These small turbines that are connected straight to the residential property are subject to an equal level of regulation and inspection as large wind farms that power a much larger area and have a much more significant impact on the environmental landscape. FERC, along with regulating new projects, interferes with the expanding of projects that are already in place. The process by which one has to go through to install a small-wind project begins with notification and pre-filing consultation with any "relevant Federal, State, and interstate resource agencies" [20]. In addition to the agencies that require consultation, Native American tribes and members of the public must be contacted.

This only marks the first step in the regulatory process, after the preliminary process is completed, a joint meeting is held for the applicable agents and members of the public to receive public comment prior to a decision being reached.

Even if FERC approval is likely, projects face the reality of additional regulatory requirements from state and local governments. One source of the development of those regulations is the National Renewable Energy Laboratory.

The NREL is a national program designed to focus on pushing the limits of renewable energy. It is a branch of the U.S. Department of Energy, Office of Energy

Efficiency and Renewable Energy, and operated by the Alliance for Sustainable Energy LLC. This program states that they want to create opportunities for job creation and land lease payments, but their purpose is also to "create a new responsibility on the part of local governments to ensure that ordinances will be established to aid the development of safe facilities that will be embraced by the community" [21]. The NREL is actively seeking to create new regulation in states and counties where few exist. Since the overview given by NREL, they have successfully lobbied for more regulation and ordinances across the nation.

There are a variety of ordinances which NREL advocates for, that fall under three themes; permission, placement, and construction. For permission, ordinances require local governments to issue permits for wind energy developments as well as ordinances that demand signage indicating warnings, the manufacturer, the owner but strictly prohibiting advertisements or promotions. In terms of placement, wind turbines must be put out of the way to limit contact with the public and respect set backs, which exist to create space between roads, private property, buildings, and phone lines. Existing wind energy ordinances also note that projects should be placed in compliance with electrical standards and Federal Aviation Administration regulations, keeping in mind the shadows created from the blades of the turbine and the windiness of the surrounding area. Other ordinances have prioritized not placing too many turbines next to each other, citing esthetic and safety reasons. For construction, NREL acknowledges ordinances that develop rules to limit "esthetic displeasure" caused by the turbine, as well as more technical aspects of the turbine such as restrictions on the arc of the blades, the height, and how much noise the turbine can emit.

The above restrictions are found in nearly every city and county ordinance package that lists regulations on wind energy projects. The strictness of each varies depending on a variety of factors, but there are some that carry across nearly all of them. Namely, signage and appearance, color, and finish, which state that no turbine may carry an advertisement or sign on them and that they be painted a neutral, non-reflective, matte color; white or gray.

Along with NREL's list of suggested ordinances, there are also site-specific laws that could affect micro-wind projects. As listed above, the National Environmental Policy Act of 1969 (NEPA) and the National Historic Preservation Act (NHP) must be taken into account when creating and placing new wind energy projects. These acts require individuals and companies to assess the environmental impact of the project as well as any effect the project could have on locations or buildings of historical significance. As for the flora and fauna, the Endangered Species Act (ESA) and General Wildlife Consultation (GWC) may come into play depending on the chosen location and scope of the wind energy project. Micro-wind regulation encompasses almost every aspect of the project ranging from the planning stage to decommission. In order to grasp the extent to which an individual or company would be subject to these regulations, we look to two counties in the Midwest.

5. Methods

To explore the regulatory environment for micro-wind energy production, we use a case-study to examine which regulations impact the implementation of micro-wind projects. We explore the regulatory environment for a micro-wind energy project from conception to integration into the energy grid of the community, and what steps they need to take to start accessing the energy created. In choosing our case, it was necessary to identify a locale with substantial wind energy potential. With a couple large wind farms in the area, Henry County, Illinois serves as our first and primary example. With that criteria established, we take a closer look at the process of implementing micro-wind projects in the county. To verify the extent of the process in Henry County, we look at Swift County, Minnesota, another Midwestern county with high wind energy potential. In both locations, numerous regulatory bodies on the federal, state, and local level have jurisdiction. We also chose counties where cleare regulations in place have a longer history of people seeking to implement renewable energy into their residences. In both cases, the potential for wind energy use is substantial and other non-micro wind generation has been explored.

The cases are different in subtle ways, as detailed in the following sections. The regulations in Swift County are somewhat more lenient, and it is easier to obtain a turbine on the basis of county level regulation, however both counties are faced with the same federal regulatory requirements that limit development. Both cases illustrate that the current regulatory approach significantly increases the costs to entry, in terms of monetary and time costs. These realities discourage people from incorporating micro wind-power on a more widespread scale, potentially defeating the regulatory motivation for a greener energy generation and unblemished environment.

6. Exploring the counties

The regulatory process that must be navigated to set up a small or micro-turbine is nearly equal to installing a large wind farm. All requirements of FERC must be met alongside any state or local regulations. This requirement discourages individuals and communities to undertake smaller non-industrial scale projects, and as a result limits the scope and form of renewable energy in use. By looking at two Midwestern counties, we find that current regulation raises the cost of small-scale wind projects and makes them unlikely to occur. As a result, potential environmental benefits are foregone in the long run.

Henry County, Illinois serves as an important illustration of the regulatory issues. In Henry County wind energy, called Wind Energy Conversion Systems (WECS) in county regulations, even at the micro-level are heavily regulated. As a result, despite having wind potential energy that is high, with an average annual wind speed of 18.86 mph (U.S. average is 16.93 mph), the county is likely not maximizing its potential due to various regulatory requirements [22].

The county ordinances prohibit any wind-energy project to be constructed, operated, or located within Henry County without having fully complied with the regulations. The county goals for the ordinance are to "preserve the health, safety, and general welfare of the public." To begin a WECS project, the applicant must obtain approval from the Henry County Planning Commission (HCPC), a variance from the Henry County Board of Zoning Appeals for any perceived or projected for the WECS project, and an Improvement Location Permit from the HCPC, which is issued by the Zoning Administrator [23]. The initial application for the WECS Commission approved use must include:

- A project summary that includes a description of the project, which entails an approximate generating capacity, potential equipment manufacturer, type of WECS, number of turbines, generating capacity for each individual turbine, maximum height and diameter of the blades and rotors, the location of the project, and a detailed description of the applicant, intentions, and business structures (should the applicant be a business).

- Names, addresses, telephone numbers of applicant, owner, and operator as well as any participating agents and property owners adjacent to any construction related to the WECS.

- A topographic map of the area with an additional mile radius from the WECS project, with contours of not more than five-foot intervals
- A full site plan with appropriate scale (the scale has several additional stipulations listed in the ordinance).
- An additional site plan showing the location of all existing and proposed underground utility lines in the WECS project area.
- Another site plan highlighting the location of amenities such as hospitals, nursing homes, and recreational areas (golf courses, trails, parks, etc.) in the WECS area.
- An agreement to properly train all emergency service agencies (Office of Emergency management, law enforcement, EMS, and fire departments) within Henry County throughout the life of the WECS, as well as addressing safety issues that arise.
- An evacuation plan and zone that complies with local emergency service agencies.
- A projected sound emissions study and map within 8 Hz to 8 kHz for the WECS area performed by a certified sound engineer.
- A small-wind energy project may not require a special use permit if used for exclusively agricultural processes

These requirements are only for the application that gets filed to the HCPC. If a project is to meet these guidelines, they must hire several people to aid them in their quest; someone to topographically map an area of land, a site planner that has the ability to create a scale model of the entire project (this step has another set of rules which make this task even more difficult), bringing in amenities and underground utility lines, and a sound engineer to determine the disturbance level of the WECS.

Once the application has been submitted and accepted by the HCPC, the WECS project is granted a one-year window to act, after which, the application lapses and the applicant must file an extension request (further regulation on what that entails) that may be valid for up to two years. The project may not be started if the applicant has not made a $75,000 deposit into an escrow account to confirm that construction can take place. This deposit may not be used toward any other application fees that are required. If the escrow account dips below the $75,000 mark, then the application is subject to revocation or denial of renewal.

In Swift County, Minnesota, the process to move forward on wind energy projects is very similar to that of Henry County, Illinois. One must apply for Land Use Permits, Conditional Use Permits and Variances which will then be reviewed under the procedures established in the Swift County Code of Ordinances. In the application, Swift County includes mostly the same requirements except they do not require a site plan highlighting nearby hospitals, nursing homes and recreation centers as well as excluding the need for evacuation plans, training for emergency services, and a projected sound emissions study.

As for specific ordinances, Swift County also has fewer restrictions than Henry County. For example, Swift County does not have any spacing regulations relative to other WECS projects, but Henry County does. The county also does not require any access restrictions, which is contrary to Henry County's requirements of a locked barrier or security fence around the WECS. Ultimately, Swift County has fewer regulations on WECS projects, but the process is still long, expensive, and tedious.

There are also regulations *within* the regulations listed above. The site plan must include turbines that are below the height limit, spaced apart appropriately, and with a diameter within the allowed range. The turbines "shall also be new equipment commercially available" [23]. A single new commercial turbine generally costs $1,300,000 per megawatt, and at 2–3 MW in power, that means most turbines cost between $2–4 million dollars [24].

Safety regulations make up a significant portion of both large- and small-scale wind projects, with more attention being paid to large wind farms. A WECS non-commercial turbine may not be closer than 1.5 times the height of the individual turbine from any property boundary lines, roadways, railroad right-of-way, or overhead transmission or distribution lines. This severely limits where a turbine may be placed. For a WECS project, this means that they must buy a large swath of property to produce any amount of electricity, and for residential projects, this impedes their ability to construct any sort of wind generator. The average commercial tower in the United States stands at roughly 280 feet, which means that for a single turbine, one would need a plot of land that is not near a road, railway, or overhead lines, and that is over 420 feet in diameter [25]. For small-wind energy systems, they must be 1.1 times the total tower height away from an occupied structure on a neighboring property and 80% the total tower height or more from an occupied structure measured from the base of the turbine. Small-wind may be located in any zoning map district with both special use and building permits, which require similar application processes as listed above [26].

Failure to adhere to the specified requirements, or violating any of the above may result in a fine of $500 per week if the offense continues without being corrected. If multiple offenses are committed, an additional $500 per week may be assessed per violation.

7. Conclusion

There is a push in congress to deregulate electricity, to open the market to allow for people to choose what their power source will be and allow a wider set of energy production to compete [27]. However, simply allowing consumers more choice in their energy source is incomplete if the regulatory environment stymies the development of innovative generation by insisting on precautionary approaches that treat all projects the same regardless of size, scope or risk. Doing so can only result in unnecessarily high costs to development which will be passed to consumers and as a result reduce the likelihood of them choosing renewable energy.

A London Economics report presented to the Department of International Development and the World Bank included an outline for how to best regulate renewable energy [28]. They suggest that to effectively allow for energy production and innovation:

- Regulation be free from political interference and promote healthy competition.

- Regulation should be maintained at a constant level and not subject to wild fluctuations, and include clear and transparent stipulations.

- Regulation should be suitable for both the cost of the project as well as the financial ability of the applicant and their party.
- Regulation should be formatted to not promote "rent-seeking behavior" by officials and federal agencies.
- Regulation should encourage incentives for developers to ensure consumers' needs are met in a satisfactory manner.
- Requirements for safety and quality must be enforced to shield both developers and consumers.

The current regulatory minefield one must navigate in any attempt to implement even micro-wind energy does not fit the mold presented in the bullet points highlighted above. In the current energy landscape, the likelihood that micro-wind will be developed further is modest and will remain modest despite specific policies that claim to incentivize renewable energy. These include Renewable Electricity Production Tax Credit, Investment Tax Credit, Residential Energy Credit, and the Modified Accelerated Cost-Recovery System [29]. The stark reality is that federal regulations as well as increasing regulation by state and local governments take many renewable energy projects untenable. As a result, many of the NREL programs which are intended to create avenues for increased renewable energy production instead act to increase the costs and barriers to entry in the renewable energy field, which discourages individuals and companies from entertaining the idea of utilizing the emerging technology.

The increased output of wind energy and small-scale projects have substantial potential to be beneficial in the long-run for more small residential communities as well as more isolated rural communities especially farms. However, if the United States wants to see a sustained increase in renewable energy, one of the simplest and easiest paths to this end is to decrease regulation, specifically by simplifying licensing requirements and regulation. It is difficult to justify the high barriers to entry for small-wind projects because they do not cause the same disturbance that a large-scale project does. They do not utilize the same infrastructure, are more flexible in their deployment, and create fewer negative externalities than larger projects. Those interested in promoting renewable energy would do well to consider the effect regulations have on preventing innovative energy solutions like micro-wind.

Author details

Ryan M. Yonk[1]*, Corbin Clark[2] and Jessica Rood[3]

1 American Institute for Economic Research, Great Barrington, USA

2 North Dakota State University, Fargo, USA

3 Truman State University, Kirksville, USA

*Address all correspondence to: ryan.yonk@aier.org

References

[1] Yonk R, Lofthouse J, Hansen M. The reality of American energy: The hidden costs of electricity policy. Westport: Praeger, 2017. 199 p.

[2] Hansen M, Simmons R, Yonk R. The regulatory noose: Logan City's adventures in micro-hydropower. Energies. 2016;9(7):482. https://doi.org/10.3390/en9070482

[3] Tullock G. The welfare costs of tariffs, monopolies, and theft. Western Economic Journal. 1967;5(3):224-232. https://doi.org/10.1111/j.1465-7295.1967.tb01923.x

[4] Stigler G. The theory of economic regulation. The Bell Journal of Economics and Management Science. 1971;2(1):3-21. https://doi.org/10.2307/3003160

[5] Arthur WB. Competing technologies, increasing returns, and lock-in by historical events. The Economic Journal. 1989;99(394)116-131. https://doi.org/10.2307/2234208

[6] North D. Institutions, institutional change and economic performance. Cambridge: Cambridge University Press, 1990. https://doi.org/10.1017/CBO9780511808678

[7] Foxon T. Technological and institutional 'lock-in' as a barrier to sustainable innovation. Imperial College Centre for Energy Policy and Technology. 2002. https://www.imperial.ac.uk/media/imperial-college/research-centres-and-groups/icept/7294726.PDF

[8] Pierson P. Increasing returns, path dependence, and the study of politics. The American Political Science Review. 2000;94(2):251-267. https://doi.org/10.2307/2586011

[9] McLaughlin P, Williams R. The consequences of regulatory accumulation and a proposed solution. Mercatus Center Working Paper. 2014. https://www.mercatus.org/system/files/McLaughlin_RegulatoryAccumulation_v1.pdf

[10] Yonk R, Simmons R, Steed B. Green vs. Green. New York: Routledge, 2013. 230 p.

[11] Wind explained [Internet]. 2021. Available from https://www.eia.gov/energyexplained/wind/electricity-generation-from-wind.php [Accessed 2021-07-15].

[12] How do wind turbines work? [Internet]. Available from https://www.energy.gov/eere/wind/how-do-wind-turbines-work [Accessed 2021-07-15].

[13] Small wind electric systems [Internet]. Available from https://www.energy.gov/energysaver/save-electricity-and-fuel/buying-and-making-electricity/small-wind-electric-systems [Accessed 2021-07-15].

[14] Small wind background information [Internet]. Available from http://www.esru.strath.ac.uk/EandE/Web_sites/09-10/Hybrid_systems/wind.htm [Accessed 2021-07-15].

[15] Micro wind turbines [Internet]. Available from https://drawdown.org/solutions/micro-wind-turbines [Accessed 2021-07-15].

[16] Wind turbine sound [Internet]. Available from https://windexchange.energy.gov/projects/sound [Accessed 2021-07-15].

[17] Planning a small wind electric system [Internet]. Available from https://www.energy.gov/energysaver/planning-small-wind-electric-system [Accessed 2021-07-15].

[18] "Energy Policy Act of 2005," Washington, DC, 2005. https://www.

congress.gov/109/plaws/publ58/PLAW-109publ58.pdf

[19] Federal energy regulatory commission [Internet]. Available from https://www.ferc.gov/ [Accessed 2021-07-15].

[20] Electronic code of federal regulations [Internet]. Available from https://www.ecfr.gov/cgi-bin/searchECFR?gp=1&ob=R&mc=true&ECFRqueryRule=%28%22WIND+ENERGY%22%29&h=2&SID=8bbaec55e2d1ba8aa076900aebd77fca&ECFRmaxHits=50&wt=ECFR_wrapper.html&host=A2&mc=true [Accessed 2021-07-15].

[21] An overview of existing wind energy ordinances [Internet]. Available from https://www.nrel.gov/docs/fy09osti/44439.pdf [Accessed 2021-07-15].

[22] Henry County Weather [Internet]. Available from http://www.usa.com/henry-county-il-weather.htm [Accessed 2021-07-15].

[23] A resolution initiating proposal to amend the Henry County code [Internet]. Available from http://www.henryco.net/attachments/UtilitiesCodeOrdanance061318.pdf [Accessed 2021-07-15].

[24] Wind turbine cost: How much? Are they worth it in 2020? [Internet]. Available from https://weatherguardwind.com/how-much-does-wind-turbine-cost-worth-it/ [Accessed 2021-07-15].

[25] Wind turbine heights and capacities have increased over the past decade [Internet]. Available from https://www.eia.gov/todayinenergy/detail.php?id=33912 [Accessed 2021-07-15].

[26] Henry County zoning ordinance appendix B wind energy [Internet]. Available from https://www.henrycty.com/Portals/0/Documents/Zoning&Building/windzoningordinance.pdf [Accessed 2021-07-15].

[27] Will renewable power prosper in a deregulated industry? [Internet]. Available from http://www.sric.org/workbook/features/V24_3/71.php [Accessed 2021-07-15].

[28] Khennas S, Barnett A. best practices for sustainable development of micro hydro power in developing countries. ESMAP Technical Paper. 2000;6. https://practicalaction.org/docs/energy/bestpractsynthe.pdf

[29] Renewable energy explained [Internet]. Available from https://www.eia.gov/energyexplained/renewable-sources/incentives.php#:~:text=The%20federal%20tax%20incentives%2C%20or,%2DRecovery%20System%20(MACRS) [Accessed 2021-07-15].